Springer Textbooks in Earth Sciences, Geography and Environment

The Springer Textbooks series publishes a broad portfolio of textbooks on Earth Sciences, Geography and Environmental Science. Springer textbooks provide comprehensive introductions as well as in-depth knowledge for advanced studies. A clear, reader-friendly layout and features such as end-of-chapter summaries, work examples, exercises, and glossaries help the reader to access the subject. Springer textbooks are essential for students, researchers and applied scientists.

More information about this series at https://link.springer.com/bookseries/15201

Geir Evensen · Femke C. Vossepoel ·
Peter Jan van Leeuwen

Data Assimilation Fundamentals

A Unified Formulation of the State and Parameter Estimation Problem

 Springer

Geir Evensen
Energy
NORCE-Norwegian Research Center
Bergen, Norway

NERSC-Nansen Environmental and Remote
Sensing Center
Bergen, Norway

Peter Jan van Leeuwen
Department of Amospheric Sciences
Colorado State University
Fort Collins, CO, USA

Meteorology Department
University of Reading
Earley Gate, Reading, UK

Femke C. Vossepoel
Department of Geoscience and Engineering
Delft University of Technology
Delft, The Netherlands

Norwegian Research Council, Delft Technology Fellowship

ISSN 2510-1307 ISSN 2510-1315 (electronic)
Springer Textbooks in Earth Sciences, Geography and Environment
ISBN 978-3-030-96711-6 ISBN 978-3-030-96709-3 (eBook)
https://doi.org/10.1007/978-3-030-96709-3

This Springer imprint is published by the registered company Springer Nature Switzerland AG
The registered company address is: Gewerbestrasse 11, 6330 Cham, Switzerland

Preface

This book summarizes about 75 years of the authors' accumulated experience within the formulation, development, and practical use of advanced data-assimilation methods. We realize that no available texts discuss the multitude of data-assimilation methods and problems in a "unified" manner and with a unified notation. Thus, we believe this book will serve as an essential resource for anyone working or planning to work in data assimilation.

We also believe this book is very suitable for an advanced course in data assimilation. The mathematical level is modest, and we explain all derivations in quite some detail. Furthermore, the book connects and gives a nearly complete overview and introduction to today's most popular data-assimilation methods.

Accompanied with the book, we are hosting a Github repository https://github.com/geirev/Data-Assimilation-Fundamentals. Here, we will place slides suitable for teaching the different chapters and include links to any YouTube video lectures. In addition, the repository allows for commenting and raising issues regarding the book's content. Thus, we encourage readers to report any errors in grammar, equations, or the general discussion, by raising an issue at this repository. It is also possible to have general discussions on the book's topics here.

During the work with this text, Geir Evensen received support from the Research Council of Norway and the companies AkerBP, Wintershall–DEA, Vår Energy, Petrobras, Equinor, Lundin, and Neptune Energy, through the Petromaks–2 DIGIRES research project (280473) (http://digires.no). He is grateful to all his data-assimilation colleagues at NORCE and NERSC in Bergen and his international collaborators. They have contributed to a long and flourishing journey within the data-assimilation research field.

Femke Vossepoel was supported by a Delft Technology Fellowship of the Delft University of Technology. She is thankful to the colleagues in her data-assimilation group in the Geoscience and Engineering department of the Delft University of Technology for many fruitful discussions, particularly to Arundhuti Banerjee for her contribution to Chap. 19. She thanks her international collaborators in data assimilation for open and inclusive dialogues.

Peter Jan van Leeuwen was supported via an Advanced Investigator Grant from the European Research Council via the CUNDA project. He is highly grateful for all teachings from other members of the Data Assimilation Research Center

(DARC) at the University of Reading, UK, his new data-assimilation group at Colorado State University, and many colleagues from all over the world.

In present-day science, where, unfortunately, harsh competition and favoritism are gaining ground, the data-assimilation community is mature and open to controversial ideas and reflective discussions, which is a true blessing.

Bergen, Norway   Geir Evensen
Delft, The Netherlands   Femke C. Vossepoel
Fort Collins, USA   Peter Jan van Leeuwen
December 2021

Contents

Symbols

Greek Symbols

α_i	Step lengths in ESMDA Eq. (8.35)
β	Constant in the Lorenz (1963) model Eq. (15.3)
β	Nonlinearity parameter in Eq. (18.1)
β	Parameter in Roessler model Eq. (12.3)
γ	Steplength in Gauss-Newton iterations Eq. (8.17)
γ	Steplength in linear solver Eq. (5.58)
γ	Steplength in particle flow iterations Eq. (18.3)
γ^i	Steplength in Gauss-Newton iterations Eq. (3.14)
$\delta(\cdot)$	The Dirac delta function Eq. (2.41)
ϵ	Constant Eq. (9.47)
ϵ	Vector of measurement errors Eq. (3.6)
ϵ_j	Vector of measurement errors Eq. (9.16)
ζ	Rate-and-state dependent friction parameter in Eq. (19.1)
η	Innovation in iterative methods Eqs. (3.22), (4.30), (5.6), and (5.16)
η	Right-hand side of linear system Eq. (5.55)
θ	Dependent model variables in Eq. (19.1)
θ	Function used in convolutiuons Eq. (5.69)
$\boldsymbol{\theta}$	Vector of uncertain model parameters Eq. (2.2)
$\boldsymbol{\theta}'$	Errors in prior vector of uncertain model parameters Eq. (2.2)
$\widehat{\boldsymbol{\theta}}$	First guess vector of uncertain model parameters Eq. (2.2)
$\boldsymbol{\Theta}$	Orthogoal random matrix Eq. (8.42)
λ	Eigenvalue from characteristic equation Eq. (12.5)
$\boldsymbol{\lambda}$	Adjoint variable Eq. (4.8)
$\delta\boldsymbol{\lambda}$	Adjoint state vector increment Eq. (5.27)
$\boldsymbol{\Lambda}$	Eigen values Eq. (8.49)
μ	Mean of samples Eq. (13.1)
ξ	Spring constant in variables in Eq. (19.3)
$\boldsymbol{\xi}$	State-vector increment in iterative methods Eqs. (3.23), (4.29), (5.6), and (5.15)

Π	Projection subtracting the mean and scaling by $\sqrt{N-1}$ Eq. (8.2)
ρ	Constant in the Lorenz (1963) model Eq. (15.2)
ρ	Gradient or residual in Eq. (5.56)
$\rho_{m \times m}$	Tapering matrix used in Kalman gain localization Eq. (10.21)
$\rho_{n \times m}$	Tapering matrix used in Kalman gain localization Eq. (10.21)
$\rho_{n \times n}$	Tapering matrix used in covariance localization Eq. (10.18)
σ	Constant in the Lorenz (1963) model Eq. (15.1)
σ	Standard deviation of samples Eq. (13.1)
σ_i	Standard deviation of state variable i Eq. (20.6)
$\mathbf{\Sigma}$	Singular values of SVD Eq. (8.51)
$\overline{\mathbf{\Sigma}}$	Updated covariance matrix Eq. (9.17)
$\widetilde{\mathbf{\Sigma}}$	Updated covariance matrix using $\widetilde{\mathbf{K}}$ Eq. (9.24)
τ	Time scale Eq. (5.59)
$\mathbf{\Upsilon}$	Ensemble of model-predicted measurements Eq. (8.7)
$\phi(\mathbf{b})$	Cost function in Eq. (5.56)
ψ_1	Stream function layer one Eqs. (20.3) and (20.4)
ψ_2	Stream function layer two Eqs. (20.3) and (20.4)
ψ	The field $\psi = \mathbf{Sv}$ in Eq. (5.53)
Ω	Relation between inital and the ith iteration ensemble anomalies Eq. (8.25)

Math Symbols

1	Vector with all elements equal to one Eq. (8.2)
a	Parameter in Roessler model Eq. (12.2)
A	Horzontal diffusion in the QG model Eq. (20.4)
$A(z)$	Vertical diffusion coefficient in Ekman-flow model Eq. (17.1)
$\mathbf{A}$	Ensemble of state-vectors anomalies Eq. (8.3)
b	Parameter in Roessler model Eq. (12.3)
$\mathbf{b}$	Vector of representer coefficients Eqs. (5.26) and (6.40)
$\mathbf{B}$	A ensemble matrix of representer coefficients Eq. (10.4)
$\mathbf{B}$	A positive definite matrix in descent methods Eq. (9.47)
$\mathbf{B}^i$	A "normalization matrix" used in iterative Gauss-Newton methods Eq. (3.14)
c	Parameter in Roessler model Eq. (12.3)
c_d	Wind-drag coeficient in the Ekman-flow model Eq. (17.4)
$c_s(\cdot)$	Function in Eq. (9.37)
$\mathbf{c}$	The field $\mathbf{c} = \mathbf{Rv}$ in Eq. (5.51)
$\mathbf{C}$	Covariance in Kernel in particle flow Eq. (18.4)
$\mathbf{C}$	Matrix in linear system Eq. (5.55)
$\mathbf{C}$	Matrix inverted in Eq. (8.38)
$\mathbf{C}_{dd}$	Measurement error covariance matrix Eq. (3.5)
$\overline{\mathbf{C}}_{dd}$	Ensemble measurement error covariance matrix Eq. (8.6)
$\mathbf{C}_{ff}$	Covariance matrix of the error in the flow map Eq. (9.28)

$\mathbf{C}_{yz}$	Covariance matrix of $\mathbf{y}$ and $\mathbf{z}$ Eq. (7.10)
$\overline{\mathbf{C}}_{yz}$	Ensemble covariance matrix of $\mathbf{y}$ and $\mathbf{z}$ Eq. (8.20)
$\mathbf{C}_{zy}$	Covariance matrix of $\mathbf{z}$ and $\mathbf{y}$ Eq. (6.41)
$\mathbf{C}_{zz}$	State covariance matrix Eq. (3.6)
$\overline{\mathbf{C}}_{zz}$	Ensemble state covariance matrix Eq. (8.4)
$\mathbf{C}_{qq}$	Model error covariance matrix Eqs. (2.25) and (3.7)
$\mathbf{C}_{xx}$	Model-state covariance matrix Eq. (2.28)
d	Spatial dimension Eq. (5.72)
$\mathbf{d}$	Measurements within an assimilation window Eq. (2.12)
$\mathbf{d}_l$	Measurements within assimilation window l Sect. 2.1
$\mathbf{D}$	Ensemble of perturbed measurements Eq. (8.5)
$\tilde{\mathbf{D}}$	Innovation term in subspace EnRML method Eq. (8.27)
$\mathcal{D}$	The measurements over all assimilation windows Eq. (2.6)
$\mathbf{E}$	Ensemble of measurement perturbations Eq. (8.5)
$\mathcal{E}$	The measurement errors over all assimilation windows Eq. (2.6)
f	Coriolis parameter Eqs. (20.3) and (20.4)
$f(\cdot)$	The argument's probability density function (pdf) Sect. 2.1
F	Frequency in Eq. (19.3)
F_1	Constant in the QG model Eq. (20.3)
F_2	Constant in the QG model Eq. (20.4)
$\mathbf{g}(\cdot)$	Vector function to map model inputs to predicted measurements Eq. (2.29)
$\mathbf{G}$	Tangent linear operator of $\mathbf{g}$, e.g., Eqs. (3.18) and (6.1)
$\overline{\mathbf{G}}$	Ensemble averaged model sensitivity Eq. (8.20)
$\mathcal{G}$	The composite measurement and model operator over all measurements Eq. (2.7)
$\mathbf{h}(\cdot)$	Nonlinear measurement operator Eq. (2.29)
H	Depth of lower boundary in the Ekman-flow model Eq. (17.3)
$\mathbf{H}$	Tangent-linear operator of meaurement function Eq. (4.9)
$\mathcal{H}$	The measurement operator over all measurements Eq. (2.6)
$\mathbf{I}$	Identity matrix Eq. (3.14)
$\mathbf{I}_N$	Identity matrix of dimension N Eq. (8.48)
$\mathcal{J}(\cdot)$	Cost function of its argument Eq. (3.2)
$k(\cdot,\cdot)$	Diagonal elements of the kernel $\mathcal{K}$ Eq. (9.50)
$k_{ii}(\cdot,\cdot)$	Diagonal kernel elements Eq. (20.6)
$\mathbf{k}$	Vertical unit vector in Ekman-flow model Eq. (17.1)
K	Number of time steps in an assimilation window Sect. 2.1
$\mathbf{K}$	Kalman gain matrix Eq. (6.38)
$\mathbf{K}$	Kernel in particle flow Eq. (18.4)
$\overline{\mathbf{K}}$	Kalman gain from prior ensemble Eq. (9.17)
$\tilde{\mathbf{K}}$	Kalman gain with model error covariance Eq. (9.23)
$\mathcal{K}$	Kernel in reproducing kernel Hilbert-space Eq. (9.43)
L	Number of assimilation windows Sect. 2.1
$\mathcal{L}(\cdot)$	Lagrangian function of its argument Eq. (4.8)

m	Number of measurements
m_i	Component of model operator Eq. (2.25)
$\mathbf{m}$	Model operator Eq. (2.24)
$\mathbf{M}$	Tangent linear model of $\mathbf{m}(\mathbf{x})$ Eqs. (2.26) and (6.3)
n	Dimension of the state vector
N	Number of ensemble members or particles
$\mathcal{N}(\cdot,\cdot)$	Normal distribution Eq. (3.6)
$\mathcal{N}(\cdot,\cdot,\cdot)$	Distribution for Gaussian pseudo-random-curves Eq. (13.1)
p_1	Potential vorticity in layer one Eq. (20.3)
p_2	Potential vorticity in layer two Eq. (20.4)
q	Scalar model error Eqs. (18.1) and (18.2)
$q(\cdot)$	Proposal density function (pdf) Eq. (9.7)
q	Errors in first guess of stochastic model errors over an assimilation window Eq. (2.4)
q	Stochastic model errors over an assimilation window Eq. (2.4)
q_k	Stochastic model error at time step t_k Eq. (2.5)
dq	Stochastic model error increment Eq. (2.24) and (9.27)
$\mathbf{Q}$	Eigen vectors Eqs. (8.39) and (8.49)
$\mathcal{Q}$	The model error over all assimilation windows Sect. 2.1
L	Length scale Eq. (5.59)
r_d	Decorrelation length of samples Eq. (13.1)
$\mathbf{R}$	Matrix of representer vectors Eq. (5.26)
$\mathcal{R}$	Representer matrix Eq. (5.44)
s	Pseudo time Eq. (9.27)
s	Pseudo time Eq. (5.69)
$\mathbf{S}$	Ensemble of possibly scaled predicted measurement anomalies Eqs. (8.22) and (8.46)
$\mathbf{S}$	Matrix of adjoint representer vectors Eq. (5.27)
t	Time
t_k	A time step
u	Advection velocity Chap. 14
u	Dependent model variables in Eq. (19.2)
u_a	Atmosperic windspeed's x-component in the Ekman-flow model Eq. (17.3)
$\mathbf{u}$	Dependent variable (two-D velocity field) in Ekman-flow model Eq. (17.1)
$\mathbf{u}$	Errors in first guess of stochastic model controls over an assimilation window Eq. (2.3)
$\mathbf{u}$	Stochastic model controls over an assimilation window Eq. (2.3)
$\mathbf{u}_a$	Atmosperic windspeed's the Ekman-flow model Sect. 17.2
$\mathbf{u}_k$	Stochastic model control at time step t_k Eq. (2.5)
$\hat{\mathbf{u}}$	First guess of stochastic model controls over an assimilation window Eq. (2.3)
$\mathbf{U}$	Left singular vectors of SVD Eq. (8.51)

$\mathcal{U}$	The model controls over all assimilation windows Sect. 2.1
v	Dependent model variables in Eq. (19.3)
v_a	Atmosperic windspeed's y-component in the Ekman-flow model Eq. (17.3)
$\mathbf{v}$	Vector used in representer method Eq. (5.53)
$\mathbf{V}$	Right singular vectors of SVD Eq. (8.51)
$\mathbf{V}$	Transform matrix Eq. (4.40)
w_j	Weights in the particle filters Eq. (9.4)
$\mathbf{w}$	Transform state variable Eq. (8.9)
$\mathbf{w}$	Transform state variable Eq. (4.40)
$\mathbf{W}$	Ensemble of transform state variables Eq. (8.10)
x	Dependent variable in Roessler model Eq. (12.1)
x	Dependent variable of the Lorenz (1963) model Eqs. (15.1–15.3)
x	Independent spatial variable in one–three dimensions Eq. (5.59)
x	Scalar model input Eqs. (18.1) and (18.2)
x_i	Component of model state vector Eq. (2.25)
$\mathbf{x}$	Model state vector over an assimilation window Sect. 2.1
$\mathbf{x}_0'$	Error in first guess of initial condition for dynamical model Eq. (2.1)
$\mathbf{x}_0$	Initial condition for dynamical model Eq. (2.1)
x_l	Model state vector over assimilation window l Sect. 2.1
$\hat{\mathbf{x}}_0$	First guess of initial condition for dynamical model Eq. (2.1)
$\delta\mathbf{x}$	State vector increment Eq. (5.17)
$d\mathbf{x}$	State vector increment Eq. (2.24)
$\mathcal{X}$	The model state over all assimilation windows Sect. 2.1
y	Dependent variable in Roessler model Eq. (12.2)
y	Dependent variable of the Lorenz (1963) model Eqs. (15.1–15.3)
y	Scalar model output Eqs. (18.1) and (18.2)
$\mathbf{Y}$	Ensemble of model-predicted measurement anomalies Eq. (8.8)
z	Dependent variable in Roessler model Eq. (12.3)
z	Dependent variable of the Lorenz (1963) model Eqs. (15.1—15.3)
z	Vertical depth as independent variable in the Ekman-flow model Eq. (17.1)
$\mathbf{z}$	State vector over an assimilation window Sect. 2.1
$\mathbf{z}_l$	State vector over assimilation window l Sect. 2.1
$\delta\mathbf{z}$	State vector increment Eqs. (3.12), (3.20), (5.5)
$\mathbf{Z}$	Ensemble of state vectors Eq. (8.1)
$\mathcal{Z}$	The state vector over all assimilation windows Sect. 2.1

Other Symbol

| $\Re$ | The space of real numbers, often raised to a dimension like $\Re^{m\times n}$ to denote a matrix |

List of Approximations

Introduction

1

Data assimilation combines prior information from numerical model simulations with observed data to obtain the best possible description of a dynamical system and its uncertainty. The purpose of using data assimilation is often to compute the best possible estimate of the model state. Alternatively, we use data assimilation to estimate the model parameters or infer the best characterization of the model forcing or controls. In some cases, we would like to find the best descriptions of combinations of uncertain state variables, parameters, and model controls, or all of them together.

Data assimilation provides the best tool for optimally combining all available information whenever one has access to a numerical model and observations of a dynamical system. The notion of *data assimilation* finds its origin in numerical weather prediction and operational oceanography. However, its mathematical formulation originates from Bayesian inference, control theory, and variational calculus. Data-assimilation methods have evolved over the three previous decades from simplistic and ad-hoc approaches to advanced techniques for sampling the Bayesian posterior. Furthermore, it is common to use similar data-assimilation methods both for state and parameter estimation. Data-assimilation practices have also spread from operational systems in the ocean and weather-prediction communities to a wide range of research fields, particularly within the geosciences and the medical, economic, transportation, chemical, biological, physical, and general statistical research. There are currently many different data-assimilation methods to choose from, and there are several routes of deriving them.

This book's *significant contribution* is the unified derivation of data-assimilation methods from a common fundamental and optimal starting point, namely Bayes' theorem. And, Bayes' theorem is indeed the optimal starting point, as Bui-Thanh (2021) pointed out. By reviewing earlier research, they show how Bayes' formula has a firm foundation within optimization. E.g., Bayes' formula arises from the joint minimization of the Kullback-Leibler (KL) divergence between a posterior and prior distribution and the mean-square errors of the data represented by the likelihood. As stated by Bui-Thanh (2021), "the first-order optimality condition of this optimization

© The Author(s) 2022
G. Evensen et al., *Data Assimilation Fundamentals*,
Springer Textbooks in Earth Sciences, Geography and Environment,
https://doi.org/10.1007/978-3-030-96709-3_1

problem is precisely Bayes' formula, and its unique updated distribution is the Bayes' posterior." But perhaps a more compelling argument is that Bayes' theorem is the natural learning framework as it elegantly shows how to update prior information when new information becomes available. One of the strengths of Bayes' theorem is that it does not try to solve the ill-defined problem of "inverting observations" but instead updates prior knowledge. In that sense, it is one of the very foundations of science.

Unique in this book is the "top-down" derivation of the assimilation methods. We start from Bayes theorem and gradually introduce the assumptions and approximations needed to arrive at today's popular data-assimilation methods. This strategy is the opposite of most textbooks and reviews on data assimilation that typically take a bottom-up approach to derive a particular assimilation method. Examples of the bottom-up approach include, e.g., the derivation of the Kalman filter from linear estimation or control theory, the derivation of the ensemble Kalman filter as a low-rank approximation of the standard Kalman filter, and the derivation of 4DVar from variational principles. The bottom-up approach derives the assimilation methods from different mathematical principles, making it difficult to compare them. Thus, it may be unclear which assumptions we base a data-assimilation formulation on and sometimes even which problem it aspires to solve. Our top-down approach allows us to categorize data-assimilation methods based on the approximations used. This approach enables the user to choose the most suitable method for a particular problem or application. Have you ever wondered about the difference between the ensemble 4DVar and the "ensemble randomized maximum likelihood" (EnRML) methods? Do you know the differences between the ensemble smoother (ES) and the ensemble Kalman smoother (EnKS)? Would you like to understand how a particle-flow filter compares to a particle filter? In this book, we will provide clear answers to several such questions.

We will show how we can consistently derive the formulations and solution methods for recursive model-state and parameter estimation from Bayes' theorem while discussing the required assumptions and approximations. We build up towards a focus on assimilation methods that attempt to sample the Bayes' posterior pdf. Thus, we search for ensemble formulations that characterize uncertainty, such as ensemble 4DVar, ensemble Kalman filters, ensemble RML, particle filters, and particle-flow filters.

This book contains two parts, where Part I covers the theory, and Part II illustrates several data-assimilation methods applied with a range of simple models.

In Part I, we start from Bayes' formula and introduce the approximations and assumptions needed to derive the various assimilation methods. In Chap. 2, we discuss the general mathematical formulation and notation that we will use throughout the book. We introduce the concept of an assimilation window and discuss problem formulations commonly solved in data assimilation. Chapter 2 is fundamental to understanding the subsequent chapters.

In Chaps. 3–5, we derive and discuss methods that solve for the maximum a posteriori solution. These are iterative methods, typically derived from a Gauss-Newton formulation introduced in Chap. 3. After that, Chap. 4 discusses the strong-constraint 4DVar approach while Chap. 5 introduces solution methods for the weak-constraint or so-called "generalized-inverse" formulation, leading to the representer method and weak-constraint 4DVar.

Chapter 6 discusses simple methods like 3DVar and Kalman filters. These methods apply significant approximations related to either linearizations or an approximate evolution of error statistics.

Then, in Chap. 7, we introduce the concept of randomized-maximum-likelihood (RML) sampling that approximately samples the posterior pdf from Bayes' formula. In RML, we minimize an ensemble of cost functions, and we derive several popular data-assimilation methods from the RML formulation. Furthermore, we can use the assimilation methods described in the previous chapters to minimize the RML ensemble of cost functions. Thus, although these methods, by design, solve for the MAP solution, we can also use them to sample the Bayesian posterior approximately. This chapter illustrates how so-called "hybrid methods" follow naturally from Bayes' theorem.

Chapter 8 takes the RML sampling one step further by using the ensemble statistics to represent the background-error-covariance matrix and propagate error statistics forward in time. Here we derive popular and highly efficient methods such as ensemble RML and ensemble Kalman filters and smoothers.

The final methods discussed in Chap. 9, include particle filters and particle-flow filters, which aspire to solve the fully nonlinear Bayesian problem through an exact sampling of the posterior pdf. Using proposal densities, we demonstrate that we can naturally combine all the data-assimilation methods derived in earlier chapters with the particle-filter techniques, showing that we can derive the hybrid approaches directly from Bayes' theorem.

To complete the theoretical discussion in Part I, we discuss localization and inflation methods in Chap. 10. After that, Chap. 11 gives an overall assessment of all the assimilation methods discussed and the approximations used to derive them.

In Part II, we discuss the performance of different assimilation strategies in simple examples and demonstrate some real applications of data assimilation to illustrate the methods' potential. We start with a Kalman filter and extended Kalman filter application with the Roessler model in Chap. 12, demonstrating the impact of linearizations in EKF.

In Chap. 13, we discuss the properties of the linear ensemble Kalman filter update and assess and demonstrate the "sub-space inversion" method that allows computing efficient EnKF updates with large data sets and correlated measurement errors. We follow up this discussion in Chap. 14, where we illustrate sequential data assimilation with a linear advection equation, and in Chap. 15, we use different ensemble methods with the chaotic Lorenz'63 model.

In Chap. 16, we switch to apply strong-constraint 4DVar with the Lorenz'63 model. We show how a formulation with multiple data-assimilation windows allows using 4DVar with a highly nonlinear model.

We then demonstrate using the representer method for solving the weak constraint 4DVar problem in Chap. 17, where we also explain the difference between the weak- and strong-constraint assumptions.

Interesting is also the nonlinear scalar example in Chap. 18, where we examine the convergence of some advanced data-assimilation methods, including iterative ensemble smoothers. We find that only a particle-flow filter can sample the correct posterior pdf in the highly nonlinear case.

In the following Chap. 19, we use a particle filter to estimate the state and parameters in a nonlinear seismic-cycle model, followed by a particle-flow implementation with a quasi-geostrophic ocean model in Chap. 20.

Finally, we present data-assimilation applications for history matching an oil-reservoir model in Chap. 21, including control-variable estimation, and in Chap. 22, we consider joint parameter estimation and control-variable estimation for predicting the Covid-19 epidemic.

This book is complementary to the following previously published books on data assimilation. Jazwinski (1970) is a masterpiece on linear and nonlinear filtering and is still relevant today. Daley (1991) focuses on atmospheric data assimilation. Bennett (1992, 2002) explains the representer method for oceanic and atmospheric data assimilation. Kalnay (2002) discusses data assimilation in meteorology. Tarantola (2005) provides a fundamental treatment of especially variational methods, emphasizing solid-Earth problems. Fichtner (2011) and Nolet (2008) are advanced and introductory texts on seismic tomography, focusing on variational methods. Lewis et al. (2006) treat the data-assimilation problem from the least-squares perspective. Evensen (2009b) gives an extensive introduction to ensemble data assimilation. Bain and Crisan (2009) provide a mathematical foundation for stochastic filtering. Oliver et al. (2008) discuss history matching in petroleum applications. Majda and Harlim (2012) discuss ensemble filtering techniques for turbulent flows, emphasizing on low-order modeling of the filtering problem. Law et al. (2015) provide a mathematical description of the probabilistic approach. Reich and Cotter (2015) consider a general dynamic-systems approach with low-dimensional examples. Van Leeuwen et al. (2015) introduce nonlinear data assimilation focussing on particle filtering. Asch et al. (2017) present statistical, variational, and hybrid data-assimilation methods and their applications. Fletcher (2017) introduces variational and ensemble data assimilation methods and numerical methods used in meteorology.

Additionally, there are several review papers on data assimilation that discuss both the methods and their applications. Evensen (2009a) reviews ensemble Kalman filters and smoothers and introduces the combined parameter and state estimation problem.

Carrassi et al. (2018) offer a mathematical description that serves as the first guide for readers relatively new to data assimilation. It provides a comprehensive overview of data-assimilation methods starting from Bayesian theory and examples that include data assimilation for chaotic systems and non-Gaussian problems.

Van Leeuwen et al. (2019) provide an overview of particle filters for high-dimensional geoscience applications. It places the particle filter in the context of other data-assimilation methods and provides a mathematical description from a probabilistic perspective. The publication includes pseudo-code for several particle-filter algorithms. Notably, the paper discusses proposal-density particle filters, transportation particle filters, and localization in particle filters. It also presents several hybrid methods that include particle filters.

Stuart (2010) gives a broad mathematical overview and presents a common mathematical framework in a Bayesian approach starting from a continuous infinite-dimensional description. He classifies methods into three categories; maximum a posteriori probability estimators, filters, and sampling methods. He continues to discuss a wide range of inverse problems in fluid mechanics, weather prediction, oceanography (Navier-Stokes), and subsurface geophysics (Darcy).

Vetra-Carvalho et al. (2018) provide a valuable overview of all ensemble Kalman filter variants in use at that time derived from a unifying framework, including pseudo-code for efficient implementation.

Several review papers address the use of data assimilation in different applications areas, e.g., for weather prediction (Bannister, 2017; Houtekamer & Zhang, 2016), history-matching of petroleum reservoir models (Aanonsen et al., 2009; Oliver & Chen, 2011), or hydrology (Liu et al., 2012; McLaughlin, 1995). This book provides easy access to these review papers for further reading into specific topics not covered here.

We are all too well aware that this book will have its shortcomings beyond typos and other mistakes. It represents our view on the field, which might be controversial in places. We had to leave out many essential subjects to keep the book focused. E.g., we do not cover the extensive literature on preconditioning in variational methods, and we do not discuss representation errors in any depth (a topic that currently is not treated well in any book).

Finally, we have not discussed the many "hybrid" methods between variational approaches and ensemble Kalman filters and between ensemble Kalman filters and particle filters. Fortunately, the material in this book should be sufficient for understanding the pros and cons of these hybrid methods. More important is perhaps our choice of references. This choice is biased by our knowledge, familiarity, and bias, and we apologize beforehand for the many omissions. We hope our friendships will not be affected and encourage colleagues to point us to gross errors.

Part I
Mathematical Formulation

The book's first part gives a systematic introduction to data-assimilation methods and formulations starting from Bayes' theorem.

Problem Formulation

2

This chapter introduces the model-state- and parameter estimation problem from basic principles starting with Bayes' theorem. We define the general problem formulation and introduce the concept of Bayes' theorem solved recursively over a sequence of assimilation time windows. We also present different assimilation- and parameter-estimation problems, including model controls and errors, and show how they fit into a similar and general framework. Furthermore, this chapter introduces the notation and problem formulation considered in the following chapters.

2.1 Bayesian Formulation

This section introduces the concept of a data assimilation window, followed by the dynamical model and its uncertain quantities, the definitions of the model state, and the state vector. Then we discuss the measurements and the measurement equation before we formulate the general Bayesian data-assimilation problem.

2.1.1 Assimilation Windows

We typically solve the data-assimilation problem sequentially over a sequence of *assimilation time windows*. We adopt a rather general definition for an assimilation window to allow for various data-assimilation formulations and methods. For some methods, the assimilation windows are fixed-length intervals in time. E.g., in atmospheric data assimilation, it is common to use assimilation windows of six or twelve hours in length. In contrast, the assimilation window is the time interval between available measurements for Kalman-filter-type methods. We will later see that some assimilation methods update the model solution over the whole window, while others compute it at a particular time. Additionally, some assimilation methods treat each assimilation window independently, while others propagate information from one window to the next.

© The Author(s) 2022
G. Evensen et al., *Data Assimilation Fundamentals*,
Springer Textbooks in Earth Sciences, Geography and Environment,
https://doi.org/10.1007/978-3-030-96709-3_2

2.1.2 Model with Uncertain Inputs

We assume a forward model that describes a dynamic process with uncertainty over an assimilation window

$$\mathbf{x}_0 = \hat{\mathbf{x}}_0 + \mathbf{x}'_0, \tag{2.1}$$

$$\boldsymbol{\theta} = \hat{\boldsymbol{\theta}} + \boldsymbol{\theta}', \tag{2.2}$$

$$\mathbf{u} = \hat{\mathbf{u}} + \mathbf{u}', \tag{2.3}$$

$$\mathbf{q} = 0 + \mathbf{q}', \tag{2.4}$$

$$\mathbf{x}_k = \mathbf{m}(\mathbf{x}_{k-1}, \boldsymbol{\theta}, \mathbf{u}_k, \mathbf{q}_k). \tag{2.5}$$

Here, $\mathbf{x}_0$ is a vector containing the model's *uncertain initial conditions* $\hat{\mathbf{x}}_0$ with uncertainty represented by $\mathbf{x}'_0$. The vector $\boldsymbol{\theta}$ denotes a set of *uncertain model parameters* with prior values $\hat{\boldsymbol{\theta}}$ and uncertainty $\boldsymbol{\theta}'$. We assume parameters $\boldsymbol{\theta}$ to be constant in time. Furthermore, we define the *uncertain model errors* $\mathbf{q}^{\mathrm{T}} = (\mathbf{q}_1^{\mathrm{T}}, \ldots, \mathbf{q}_K^{\mathrm{T}})$ with uncertainty $\mathbf{q}'$. The model errors account for missing physics in the model equations and numerical discretization errors. Note that we distinguish between the time-independent parameter uncertainty and time-dependent model errors by defining $\boldsymbol{\theta}$ and $\mathbf{q}$ separately. The *uncertain model controls* $\mathbf{u}^{\mathrm{T}} = (\mathbf{u}_1^{\mathrm{T}}, \ldots, \mathbf{u}_K^{\mathrm{T}})$ with uncertainty $\mathbf{u}'$ represent various forms of time-dependent but uncertain model forcing. We have defined K as the number of time steps over the assimilation window. For simplicity, we have ignored any boundary conditions and their potential uncertainty, as this would add additional constraints to the model system.

2.1.3 Model State

We define $\mathbf{x}^{\mathrm{T}} = (\mathbf{x}_0^{\mathrm{T}}, \ldots, \mathbf{x}_K^{\mathrm{T}})$ as a sequence of model-state vectors over an assimilation window. On some occasions, we will, for short, write $\mathbf{x} = \mathbf{m}(\mathbf{x}_0, \boldsymbol{\theta}, \mathbf{u}, \mathbf{q})$ where the model operator predicts the model state over the whole assimilation window. We differentiate between the model state $\mathbf{x}$ and the data-assimilation problem's more general state vector, discussed next, although they will be the same in some cases.

2.1.4 State Vector

We define the data-assimilation problem's state vector $\mathbf{z}$, containing all the uncertain quantities we wish to estimate. The variables included in $\mathbf{z}$ depend on the data-assimilation problem at hand and its formulation. There are, however, two main formulations to be aware of.

In the first formulation, $\mathbf{z}$ includes the model prediction, or model state, $\mathbf{x}$, or a subset of $\mathbf{x}$ (e.g., $\mathbf{x}_K$). In this case, we update the model state $\mathbf{x}$ directly, and we denote it as the *model-state formulation*, where we can have $\mathbf{z}^{\mathrm{T}} = (\mathbf{x}^{\mathrm{T}}, \boldsymbol{\theta}^{\mathrm{T}}, \mathbf{u}^{\mathrm{T}})$, excluding the model error $\mathbf{q}$.

In the second formulation, we treat the model error as an uncertain variable that we will estimate. We then write $\mathbf{z}^T = (\mathbf{x}_0^T, \boldsymbol{\theta}^T, \mathbf{u}^T, \mathbf{q}^T)$ and note that as soon as we give $\mathbf{z}$, we also determine the model state $\mathbf{x}$. We denote this formulation as the *forcing formulation* because the estimated model errors force the model.

It is important to note that we cannot simultaneously estimate both $\mathbf{x}$ and $\mathbf{q}$ as the model equations uniquely connect these variables. We will also see below that different assimilation methods will use either one or the other formulation.

2.1.5 Formulation Over Multiple Assimilation Windows

To understand which approximations we impose when solving the data-assimilation problem for a single assimilation window, we start by formulating the general or complete problem over multiple assimilation windows. The model state over L assimilation windows is the model-state trajectory $\mathcal{X}^T = (\mathbf{x}_1^T, \ldots, \mathbf{x}_L^T)$. For the remainder of the book, we will use an index k denoting a particular time step t_k, while an index l refers to an assimilation window. We also define $\mathcal{U}^T = (\mathbf{u}_1^T, \ldots, \mathbf{u}_L^T)$ and $Q^T = (\mathbf{q}_1^T, \ldots, \mathbf{q}_L^T)$ as the time sequences of controls and model errors.

We gather the model-state trajectory, $\mathcal{X}$, the parameters $\boldsymbol{\theta}$, the model controls $\mathcal{U}$, and the model errors Q, into the *state vector* in either the state formulation $\mathcal{Z}^T = (\mathcal{X}^T, \boldsymbol{\theta}^T, \mathcal{U}^T)$ or the forcing formulation $\mathcal{Z}^T = (\mathcal{X}_0^T, \boldsymbol{\theta}^T, \mathcal{U}^T, Q^T)$ where $\mathcal{X}_0$ is the initial condition for the first assimilation window. So $\mathcal{Z}$ holds all uncertain quantities over all the assimilation windows.

2.1.6 Measurements with Errors

We also have a vector of measurements $\mathcal{D}$ of the model predicted state $\mathcal{X}$ represented by a measurement equation

$$\mathcal{D} = \mathcal{H}(\mathcal{X}) + \mathcal{E}. \tag{2.6}$$

Here $\mathcal{H}$ is the so-called measurement operator, a potentially nonlinear function that maps the model state vector $\mathcal{X}$ into measurement space. The matrix $\mathcal{E}$ contains the measurement errors. Note that we will use the terms "measurement errors" and "observation errors" interchangeably.

We note that $\mathcal{Z}$ includes $\mathcal{X}$ in the state formulation, and we can equally write

$$\mathcal{D} = \mathcal{G}(\mathcal{Z}) + \mathcal{E}, \tag{2.7}$$

where $\mathcal{G}$ relates the measurements to $\mathcal{Z}$.

In the forcing formulation we can still use Eq. (2.7) but rewrite it as

$$\mathcal{D} = \mathcal{G}(\mathcal{Z}) + \mathcal{E} = \mathcal{H}\big(\mathcal{M}(\mathcal{Z})\big) + \mathcal{E}, \tag{2.8}$$

where $\mathcal{G}$ measures the model prediction from the input state vector, and $\mathcal{M}(\cdot)$ is the model operator that maps $\mathcal{Z}$ to $\mathcal{X}$.

The term $\mathcal{E}$ contains the measurement errors, including instrument errors and possibly a representation error that accounts for different representations of reality by the measurements and the model. Representation errors are typically errors due to unresolved scales and processes, and Hodyss and Nichols (2015) and Van Leeuwen (2015) provide a further understanding of their origin and how to treat them in the data assimilation problem. An example of representation error occurs in satellite-altimetry data measuring the height of the sea surface. Apart from height differences induced by large-scale ocean currents and eddies, these measurements also contain height signals resulting from ocean tides. Large-scale ocean models often do not include these tides for computational reasons, and hence the observations have processes that the ocean models do not resolve. The representation errors may also result from using an erroneous measurement operator or errors introduced during the measurement preprocessing (Janjić et al., 2018).

2.1.7 Bayesian Inference

It is convenient to formulate the data-assimilation problem as a Bayesian inference problem. This formulation is entirely general and applies to both the state and forcing formulations discussed above. The starting point is an initial or prior probability density, $f(\mathcal{Z})$, of the quantity of interest, $\mathcal{Z}$. In the following, we use the notation $f(\cdot)$ to describe the argument's probability density function (pdf), meaning that, for instance, $f(\mathcal{Z})$ denotes the pdf of $\mathcal{Z}$, and $f(\mathbf{q})$ denotes the pdf of $\mathbf{q}$.

The measurements enter the data-assimilation problem via the likelihood $f(\mathcal{D}|\mathcal{Z})$. The likelihood is one of the less well-understood parts of the data-assimilation problem. To fully grasp its meaning, we should distinguish between the actual measurement process and how we treat measurements in the data-assimilation process. During the measurement process, the measurements are random variables. We measure them from the unknown true state of the system, to which nature adds a random draw from the measurement-error pdf. *Once we have collected the measurements, they are not random but fixed.* Of course, the measurements have errors, but that does not make them random because we know them exactly. This realization means that the likelihood is not the pdf of the measurements but rather a function of the state, and the measurements are fixed in that function. Detailed further discussions can be found in Van Leeuwen (2015, 2020).

To obtain the likelihood, we need to calculate $f(\mathcal{D}|\mathcal{Z})$ for each possible $\mathcal{Z}$, which is the (unnormalized) probability that this specific vector $\mathcal{Z}$ would result in the fixed set of measurements. Note that to obtain the likelihood $f(\mathcal{D}|\mathcal{Z}) = f(\mathcal{E}) = f\big(\mathcal{D} - \mathcal{G}(\mathcal{Z})\big)$, we need to prescribe the probability-density function of the measurement errors $\mathcal{E}$.

Assuming knowledge of the probability density $f(\mathcal{Z})$ of all the uncertain variables in the state vector $\mathcal{Z}$ and the likelihood $f(\mathcal{D}|\mathcal{Z})$, we can define a general form of the data-assimilation problem through Bayes' theorem. We can derive Bayes' theorem from the definition of conditional pdfs,

$$f(\mathcal{Z}, \mathcal{D}) = f(\mathcal{Z}|\mathcal{D})f(\mathcal{D}) = f(\mathcal{D}|\mathcal{Z})f(\mathcal{Z}), \tag{2.9}$$

where we solve for $f(\mathcal{Z}|\mathcal{D})$ to get

$$f(\mathcal{Z}|\mathcal{D}) = \frac{f(\mathcal{D}|\mathcal{Z}) \, f(\mathcal{Z})}{f(\mathcal{D})}. \tag{2.10}$$

This equation becomes Bayes' theorem when used with the fixed measurement set $\mathcal{D}$. Lorenc (1988) and Tarantola (1987) introduced the Bayesian formulation for time-independent problems, and it was extended and generalized for time-dependent problems by Van Leeuwen and Evensen (1996). The pdf of the state vector $\mathcal{Z}$ given the measurements $\mathcal{D}$, i.e., $f(\mathcal{Z}|\mathcal{D})$ is the solution to the data-assimilation problem. We stress that we know the observations in data assimilation, $\mathcal{D}$ is not a random vector, and Bayes' theorem is a point-wise equation for each vector $\mathcal{Z}$.

The denominator on the right-hand side is the marginal pdf of the measurements, $f(\mathcal{D})$. It acts as a normalization constant that ensures that the posterior pdf integrates to one. Indeed, we can write

$$f(\mathcal{D}) = \int f(\mathcal{D}, \mathcal{Z}) \, d\mathcal{Z} = \int f(\mathcal{D}|\mathcal{Z}) f(\mathcal{Z}) \, d\mathcal{Z}, \tag{2.11}$$

making the normalization explicit. This normalization constant has also given rise to misunderstandings. Some have argued that $f(\mathcal{D}) = 1$ since we know the measurements, but this is incorrect. Instead, one should evaluate the pdf of the measurements at the value of the fixed measurements $\mathcal{D}$. This pdf centers on the unknown true state's measurement, so the above equation is the proper way to evaluate $f(\mathcal{D})$.

However, this normalization constant is seldom needed explicitly, and we will not use it in the rest of this book. Actually, $f(\mathcal{D})$ only shows up when evaluating different numerical models given a set of measurements via the "model evidence," which is the normalization constant assessed for each of these models, e.g., the ECMWF model versus the Met Office model. The model with the highest $f(\mathcal{D})$ is considered the best model. In contrast, a proper Bayesian would set up a prior over the different numerical models and interpret the normalization factor as the likelihood in this higher-order data-assimilation problem. Since the number of models tested is finite, this is a discrete data-assimilation problem where the state vector $\mathcal{Z}$ contains a complete numerical model.

It is essential to realize that Bayes' theorem describes a forward problem and not an inverse problem. We start with the prior probability density function and multiply it with the likelihood to form the posterior probability density function, which is the solution to the problem. Of course, many practical data-assimilation methods solve an inverse problem to find an approximation to the posterior, but many others do not. As such, we can consider inverse problem theory as a subset of Bayesian inference; see, for instance, the introduction in Van Leeuwen et al. (2015).

2.2 Recursive Bayesian Formulation

The Bayesian formulation above updates the solution over the whole assimilation period, including multiple assimilation windows, in one go. This formulation is not always convenient, as, in many data-assimilation problems, the measurements become available sequentially. Thus, we will next introduce two approximations to simplify the process of solving Eq. (2.10).

2.2.1 Markov Model

Approximation 1 (Model is 1st-order Markov process)   *We assume the dynamical model is a 1st-order Markov process.*   □

A first-order Markov process denotes that the future is independent of the past if the present is known. If the model is a 1st-order Markov process, it follows that we can compute the model solution in one assimilation window from the solution in the previous window. We can then express mathematically the condition

$$f(\mathbf{z}_l|\mathbf{z}_{l-1}, \mathbf{z}_{l-2}, \ldots, \mathbf{z}_0) = f(\mathbf{z}_l|\mathbf{z}_{l-1}), \qquad (2.12)$$

where we remind the reader that l is the index of an assimilation window.

From Approx. 1, we can use Eq. (2.12) and write $f(\mathcal{Z})$ as a recursion over the assimilation windows $l \in (1, \ldots, L)$,

$$\begin{aligned} f(\mathcal{Z}) &= f(\mathbf{z}_0) \, f(\mathbf{z}_1|\mathbf{z}_0) \, f(\mathbf{z}_2|\mathbf{z}_1) \cdots f(\mathbf{z}_L|\mathbf{z}_{L-1}) \\ &= f(\mathbf{z}_0) \prod_{l=1}^{L} f(\mathbf{z}_l|\mathbf{z}_{l-1}). \end{aligned} \qquad (2.13)$$

We notice that the assumption of the model being a Markov process affects the model-state evolution in time. However, we use the Markov property to formulate the model prior as a recursion over the assimilation windows.

2.2.2 Independent Measurements

Next, we introduce the following assumption on the measurements.

Approximation 2 (Independent measurements)   *We assume that measurements are independent between different assimilation windows.*

Independent measurements mean that their errors are uncorrelated, so we now assume the measurement-error correlations are zero between measurements in different assimilation windows. If we use Approx. 2, we can write the likelihood for the measurement vector $\mathcal{D}$ as a product of independent likelihoods, one for each assimilation window, as

$$f(\mathcal{D}|\mathcal{Z}) = \prod_{l=1}^{L} f(\mathbf{d}_l|\mathbf{z}_l). \qquad (2.14)$$

The assumption of independence of measurements collected in different assimilation windows is the first grave approximation we make, as such correlations often exist, and we still neglect them. However, we retain the possibility of having correlated measurement errors for the measurements collected within each assimilation window.

Note that, by "independent measurements," we do not mean that the measurements' information is independent. This terminology typically conveys that the *measurement errors* are uncorrelated. E.g., we can measure a temperature at a location twice using different sensors, so the measurement errors are uncorrelated. But, since they measure the same quantity, we do not double the information. As Evensen and Eikrem (2018) discussed, there is redundancy in the measurement information. Likewise, spatial correlations in the measured quantity will also result in measurements with correlated information. The measurement errors can still be uncorrelated.

2.2.3   Recursive form of Bayes'

The general form of Bayes' in Eq. (2.10) now becomes

$$f(\mathcal{Z}|\mathcal{D}) \propto \prod_{l=1}^{L} f(\mathbf{d}_l|\mathbf{z}_l)\, f(\mathbf{z}_l|\mathbf{z}_{l-1})\, f(\mathbf{z}_0). \tag{2.15}$$

By rearranging the order of the multiplications, it is possible to write Eq. (2.15) as a recursion, following Evensen & Van Leeuwen, (2000),

$$f(\mathbf{z}_1, \mathbf{z}_0|\mathbf{d}_1) = \frac{f(\mathbf{d}_1|\mathbf{z}_1)\, f(\mathbf{z}_1|\mathbf{z}_0)\, f(\mathbf{z}_0)}{f(\mathbf{d}_1)}, \tag{2.16}$$

$$f(\mathbf{z}_2, \mathbf{z}_1, \mathbf{z}_0|\mathbf{d}_1, \mathbf{d}_2) = \frac{f(\mathbf{d}_2|\mathbf{z}_2)\, f(\mathbf{z}_2|\mathbf{z}_1)\, f(\mathbf{z}_1, \mathbf{z}_0|\mathbf{d}_1)}{f(\mathbf{d}_2)}, \tag{2.17}$$

$$\vdots$$

$$f(\mathcal{Z}|\mathcal{D}) = \frac{f(\mathbf{d}_L|\mathbf{z}_L)\, f(\mathbf{z}_L|\mathbf{z}_{L-1})\, f(\mathbf{z}_{L-1}, \ldots, \mathbf{z}_0|\mathbf{d}_{L-1}, \ldots, \mathbf{d}_1)}{f(\mathbf{d}_L)}. \tag{2.18}$$

Thus, we have defined the data-assimilation problem as a recursion in time. We start in Eq. (2.16), with $f(\mathbf{z}_0)$ being the prior density for the initial conditions $\mathbf{z}_0$. Then, $f(\mathbf{z}_1|\mathbf{z}_0)$ denotes the integration of the model from the initial condition to predict the pdf of $\mathbf{z}_1$ given $\mathbf{z}_0$. The multiplication with the likelihood $f(\mathbf{d}_1|\mathbf{z}_1)$ conditions the model prediction on the measurements $\mathbf{d}_1$. The posterior estimate is then the joint pdf for $\mathbf{z}_0$ and $\mathbf{z}_1$ conditioned on the measurements $\mathbf{d}_1$, denoted by $f(\mathbf{z}_1, \mathbf{z}_0|\mathbf{d}_1)$.

The posterior estimate from Eq. (2.16) now becomes the prior in Eq. (2.17), where we again integrate the model, $f(\mathbf{z}_2|\mathbf{z}_1)$, and condition on the data, $f(\mathbf{d}_2|\mathbf{z}_2)$, to obtain the posterior $f(\mathbf{z}_2, \mathbf{z}_1, \mathbf{z}_0|\mathbf{d}_1, \mathbf{d}_2)$, which becomes the prior for the following recursion.

2.2.4   Marginal Bayes' for Filtering

When the aim is to make better predictions, we can simplify the recursion in
Eqs. (2.16)–(2.17) by exploiting the Markovian property of the model and the sequential nature of the data-assimilation problem. Thus, by integrating out the model states
of previous assimilation time windows, we can write the recursion in terms of the
marginals as

$$f(\mathbf{z}_1|\mathbf{d}_1) = \frac{f(\mathbf{d}_1|\mathbf{z}_1) \int f(\mathbf{z}_1|\mathbf{z}_0) f(\mathbf{z}_0) \, d\mathbf{z}_0}{f(\mathbf{d}_1)} = \frac{f(\mathbf{d}_1|\mathbf{z}_1) f(\mathbf{z}_1)}{f(\mathbf{d}_1)}, \qquad (2.19)$$

$$f(\mathbf{z}_2|\mathbf{d}_1, \mathbf{d}_2) = \frac{f(\mathbf{d}_2|\mathbf{z}_2) \int f(\mathbf{z}_2|\mathbf{z}_1) f(\mathbf{z}_1|\mathbf{d}_1) \, d\mathbf{z}_1}{f(\mathbf{d}_2)} = \frac{f(\mathbf{d}_2|\mathbf{z}_2) f(\mathbf{z}_2|\mathbf{d}_1)}{f(\mathbf{d}_2)}, \quad (2.20)$$

$$\vdots$$

$$f(\mathbf{z}_L|\mathcal{D}) = \frac{f(\mathbf{d}_L|\mathbf{z}_L) \int f(\mathbf{z}_L|\mathbf{z}_{L-1}) f(\mathbf{z}_{L-1}|\mathbf{d}_{L-1}, \dots, \mathbf{d}_1) \, d\mathbf{z}_{L-1}}{f(\mathbf{d}_L)}$$

$$= \frac{f(\mathbf{d}_L|\mathbf{z}_L) f(\mathbf{z}_L|\mathbf{d}_{L-1})}{f(\mathbf{d}_L)}. \qquad (2.21)$$

Note that by solving for the marginals, we are not solving the complete original
problem defined by Bayes' theorem in Eq. 2.10. We are applying the following
approximation.

> **Approximation 3** (Filtering assumption)   *We approximate the full smoother solution with a sequential data-assimilation solution. We only update the solution in the
> current assimilation window, and we do not project the measurement's information
> backward in time from one assimilation window to the previous ones.* ☐

We recursively accumulate more and more information from the measurements
from time window to time window. Hence, $\mathbf{z}_l$ contains the information from all the
previous measurements, including those from the l'th assimilation window, and is
the ideal starting point for predicting $\mathbf{z}_{l+1}$ and onwards.

We note that if the assimilation window is a single model time step, then Approx. 3
reduces to the standard filter approximation in, e.g., Kalman-filter methods, which
update the solution at the current time before continuing the integration.

Another attractive property of these equations is their similarity. We are solving
the same computational problem in each update step. We have a state vector for the
current time window containing the information from all the previous measurements,
and we combine it with the new measurements. Thus, if we define $\mathbf{z}_l$ to represent the
model prediction at time window l and $\mathbf{d}_l$ the measurements for this time window,
we can write the general update equation as

$$f(\mathbf{z}_l|\mathbf{d}_l) = \frac{f(\mathbf{d}_l|\mathbf{z}_l) f(\mathbf{z}_l)}{f(\mathbf{d}_l)}, \qquad (2.22)$$

or more generally as

$$f(\mathbf{z}|\mathbf{d}) = \frac{f(\mathbf{d}|\mathbf{z})\,f(\mathbf{z})}{f(\mathbf{d})}, \tag{2.23}$$

which is again just the Bayes formula in Eq. (2.10) but now for a subset of the state vector and the measurements.

2.3   Error Propagation

So far, we have considered formulations for updating the model solution over an assimilation window and assume that we have a prior pdf for the solution. We will now discuss how to obtain or estimate this prior distribution and propagate it to the next assimilation window.

2.3.1   Fokker–Planck Equation

Ideally, we know the full pdf at the end of the previous assimilation window or equivalently at the beginning of the current assimilation window. We can compute the evolution of the prior pdf from the start of one assimilation window to the next, $f(\mathbf{z}_l|\mathbf{z}_{l-1})$, from the Fokker–Planck equation. We start with a stochastic model with additive Gaussian model errors forming a Markov process, initialized with the model state $\mathbf{x}$ from $\mathbf{z}_l$. We assume that the prior model parameters do not change over the assimilation window, so we omit them here for ease of notation. The stochastic model equation reads

$$d\mathbf{x} = \mathbf{m}(\mathbf{x})\,dt + d\mathbf{q}, \tag{2.24}$$

giving the increment in the model state $d\mathbf{x}$ resulting from a time increment dt and the stochastic forcing $d\mathbf{q}$. From this model equation, one can derive the standard form of the Fokker–Planck equation (also named Kolmogorov's equation),

$$\frac{\partial f(\mathbf{x})}{\partial t} + \sum_i \frac{\partial\big(m_i(\mathbf{x})\,f(\mathbf{x})\big)}{\partial x_i} = \frac{1}{2}\mathbf{C}_{qq}\sum_{i,j}\frac{\partial^2 f(\mathbf{x})}{\partial x_i\,\partial x_j}, \tag{2.25}$$

which describes the time evolution of the model state vector's probability density $f(\mathbf{x})$. For the case with non-additive model errors, it becomes more elaborate to derive an equation for the probability density's time evolution, and we will not discuss that here. In Eq. (2.25), m_i is the component number i of the model operator $\mathbf{m}$, and $\mathbf{C}_{qq}$ is the model-error covariance matrix. This equation is the fundamental equation for the evolution of error statistics. The equation describes the probability change in a local volume resulting from the model dynamics and diffusion terms. The model-dynamics term is a divergence of the probability flux into the local volume. In contrast, the diffusion term tends to smooth $f(\mathbf{x})$ due to the stochastic model forcing. Unfortunately, we cannot solve this equation in high dimensions. We refer to Jazwinski (1970) for the actual derivation and further discussion.

2.3.2 Covariance Evolution Equation

By taking moments of the Fokker–Planck equation, we can derive an evolution equation for statistical moments of the uncertainty in the model state vector $\mathbf{x}$. Alternatively, we can derive an evolution equation for the error covariance matrix by comparing the evolution of the true model state with that of our estimated model state as

$$\mathbf{x}_{k+1}^t = \mathbf{m}(\mathbf{x}_k^t) + \mathbf{q}_k \approx \mathbf{m}(\mathbf{x}_k) + \mathbf{M}_k(\mathbf{x}_k^t - \mathbf{x}_k) + \mathbf{q}_k, \tag{2.26}$$

$$\mathbf{x}_{k+1} = \mathbf{m}(\mathbf{x}_k), \tag{2.27}$$

where we used a first-order Taylor expansion in the first equation, assuming that $\mathbf{x}_k^t - \mathbf{x}_k$ is small. The superscript, t, denotes "true." Note that the linearized model $\mathbf{M}_k$ can include estimates of the model parameters. By subtracting Eq. (2.27) from Eq. (2.26), multiplying the resulting equation with its transpose, and taking the expectation, we obtain the error covariance equation used in Kalman filters,

$$\mathbf{C}_{xx,k+1} \approx \mathbf{M}_k \mathbf{C}_{xx,k} \mathbf{M}_k^{\mathrm{T}} + \mathbf{C}_{qq,k}. \tag{2.28}$$

Here $\mathbf{M}_k$ is the model's tangent-linear operator evaluated at $\mathbf{x}_k$ and $\mathbf{C}_{qq}$ is the model error covariance matrix. The extended Kalman filter (EKF) also uses this linearized error-evolution equation. In addition to an immense computational load for real-size geoscience models, Evensen (1992) showed that the approximate linear equation can not saturate unstable modes of the model, and one can experience unbounded error-variance growth. See also the example in Chap. 12.

2.3.3 Ensemble Predictions

An alternative to the two previous approaches is to use a Monte-Carlo method for propagating the error statistics. For example, if we have an ensemble of samples from a pdf at time t_k, we can integrate these samples forward until t_{k+1} by running the dynamical model for each sample separately. If the model contains model errors, we can treat these by using a stochastic model description. Ensemble integration is the approach used in the ensemble methods. It involves creating a large and finite number of model realizations or samples that represent our prior understanding of the system and its uncertainties. Ensemble Kalman filters, particle filters, and particle-flow filters all use this approach. The only approximation of this approach is the limitation of the ensemble size to a finite number of model realizations or ensemble members. Thus, ensemble integration is an attractive method for propagating the uncertainty information over an assimilation window or from one assimilation window to the next. In Fig. 2.1, the ensemble integration is illustrated by the blue lines indicating the prior ensemble integration in an ensemble smoother. The figure also shows how the ensemble prediction (represented by the green lines) captures the uncertainty through the updated ensemble members' spread.

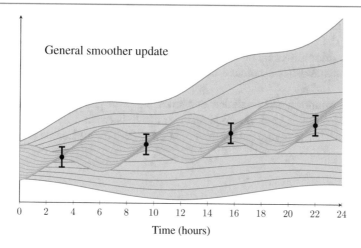

General smoother update

Time (hours)

Fig. 2.1 The figure illustrates using a general ensemble smoother (Sect. 2.4.1). The black dots denote observations with an error of one standard deviation. We first run a full ensemble integration over the whole assimilation window, indicated by blue lines in the blue envelope. After that, we update the ensemble simultaneously in space and time, resulting in the green lines in the green envelope

2.4 Various Problem Formulations

In the current formulation, the state vector $\mathbf{z}$ is rather general. We can define $\mathbf{z}$ to contain the model state over an assimilation time window or at a single instant in time. It is also possible to include other uncertainties, such as model parameters, model controls, and model errors. We will see below that the specific solution procedure will be the same in any of these cases.

2.4.1 General Smoother Formulation

A general smoother formulation solves the original Bayes' formula (2.22) in an assimilation window. Let's define the state vector $\mathbf{z} = \mathbf{x} = \mathbf{m}(\mathbf{x}_0, \mathbf{q})$ as the model solution over the whole assimilation window with distributed measurements. We also allow the model to have errors $\mathbf{q}$ but do not include uncertain parameters and controls in this example. In this case, we will update the model state $\mathbf{x}$, which connects to the model-predicted measurements $\mathbf{y}$ through the functional

$$\mathbf{y} = \mathbf{g}(\mathbf{z}) = \mathbf{h}(\mathbf{x}) = \mathbf{h}\big(\mathbf{m}(\mathbf{x}_0, \mathbf{q})\big). \tag{2.29}$$

Here the vector function $\mathbf{g}(\mathbf{z})$ is just the application of the measurement operator acting on the model prediction to generate the predicted measurements. We then compare the predicted measurements to the actual measurements in the likelihood.

We illustrate the formulation in an ensemble setting. The general smoother starts with a prior ensemble integration over the assimilation window as indicated by the blue lines in Fig. 2.1. We obtain the ensemble estimate of the posterior pdf by combining the prior ensemble in space and time with the likelihood. As such, it updates the prior ensemble and results in an updated ensemble that is closer to the measurements and has a reduced uncertainty, as illustrated by the green lines in Fig. 2.1.

For long windows and nonlinear dynamics the prior pdf becomes significantly non-Gaussian. In this case, methods that assume a Gaussian prior will struggle, as we illustrate when applying the ensemble smoother (ES) for the Lorenz (1963) model in Chap. 15.

2.4.2 Filter Formulation

Let's define $\mathbf{z} = \mathbf{x}_K$ as the model solution at the end of the assimilation window, where we assume we have measurements. We can then compute the so-called filter solution typically solved by particle filters and Kalman filter methods. In this case, we have the state $\mathbf{x}_K$ connected to the model-predicted measurements $\mathbf{y}$ through the measurement-functional $\mathbf{h}(\mathbf{x}_K)$,

$$\mathbf{y} = \mathbf{g}(\mathbf{z}) = \mathbf{h}(\mathbf{x}_K). \tag{2.30}$$

The vector function $\mathbf{g}(\mathbf{z})$ now equals the measurement operator $\mathbf{h}(\mathbf{x}_K)$, which maps the model solution at the end of the assimilation window to the predicted measurements $\mathbf{y}$. We can then compute the updated solution at the end of the assimilation window from Bayes' formula in Eq. (2.23), as illustrated in Fig. 2.2. After that, we continue the integration through the next assimilation window. Thus, we have

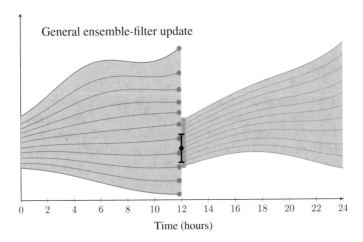

Fig. 2.2 The figure illustrates the general ensemble-filter update of an ensemble prediction (Sect. 2.4.2). The filter updates the ensemble at the measurement time before continuing the ensemble integration

integrated away the model solution from all previous time steps of the assimilation window to create the marginal pdf at the end of the assimilation window.

The filter approach can alleviate a potential practical disadvantage of the general smoother discussed in the previous section. The prior covers a large time window in the smoother-formulation and might be less accurate for measurements later in the window. Furthermore, if the model is highly nonlinear, strong non-Gaussian prior pdfs can develop. An advantage of the filter approach is that we can divide an assimilation window with observations at different time instances into several shorter windows, each ending at an observation time. This approach results in a sequential data-assimilation problem that facilitates, e.g., Kalman filters, ensemble Kalman filters, and particle filters. The more frequent model updating will keep the model closer to the observations. Furthermore, the prior at each measurement time remains relatively narrow and more Gaussian than in the smoother formulation, resulting in a sequence of more accurate updates. Several of the example chapters consider various filter applications.

2.4.3 Recursive Smoother Formulation

Evensen and Van Leeuwen (2000) introduced a recursive smoother formulation as an extension of the filtering problem. The state vector is now $\mathbf{z}^\mathrm{T} = (\ldots, \mathbf{x}_{l-1}^\mathrm{T}, \mathbf{x}_l^\mathrm{T})$ containing the model state at all, or several previous and the current assimilation windows. And we have the measurements located at the end of the current assimilation window. Thus, we have a problem where

$$\mathbf{y} = \mathbf{g}(\mathbf{z}) = \mathbf{h}(\mathbf{x}_K), \tag{2.31}$$

but we compute the update for the whole $\mathbf{z}$.

Instead of solving for the marginal in Eq. (2.20), we solve the recursion in Eq. (2.17) while processing the measurements sequentially as in the filter formulation. This approach recursively introduces the information from additional measurements in every new assimilation window and "projects" this information at previous times. The formulation inherits the advantages from the filter formulation discussed in Sect. 2.4.2. We will discuss this approach when applying the ensemble Kalman smoother in Chap. 15, and see also the Algorithm 7 in Chap. 6.

Note that the solution at the final assimilation time is identical to the filter solution. Therefore, the recursive smoother is not adding any value for prediction problems, but the formulation is an excellent alternative to the general smoother for hindcast problems (see Fig. 2.3). We can also use this formulation as a "lagged" smoother where we only update the state vector for a selected number of previous assimilation windows. We then exploit that the measurement's information decorrelates with time in nonlinear models with model errors, so, in practice, we apply a form of localization as discussed in Chap. 10. Thus, the recursive smoother introduces the measurements sequentially as in the filter, while the general smoother in Sect. 2.4.1 computes one global update over the whole assimilation window in one go.

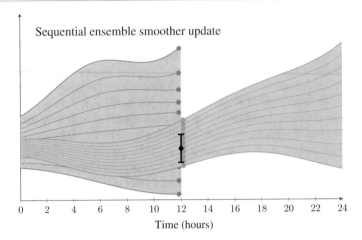

Fig. 2.3   The figure illustrates using a recursive ensemble smoother (Sect. 2.4.3). Like in the filter update, we update the ensemble at the measurement time before continuing the integration, but we also update the ensemble at all previous times

2.4.4   A Smoother Formulation for Perfect Models

An often used assimilation formulation assumes that the state vector $\mathbf{z} = \mathbf{x}_0$ only contains the model state at the beginning of the time window. Hence, we exclude uncertain parameters and controls, and we assume no model errors. We then estimate the initial model state for the assimilation window and obtain the solution over the assimilation window by integrating the model from the updated initial conditions. The equation for the predicted measurements $\mathbf{y}$ is

$$\mathbf{y} = \mathbf{g}(\mathbf{z}) = \mathbf{h}\big(\mathbf{m}(\mathbf{x}_0)\big). \tag{2.32}$$

Here we start with the solution at the beginning of the time window and integrate the model $\mathbf{x} = \mathbf{m}(\mathbf{x}_0)$ over the time window to make a prediction. Then we apply the measurement functional on the model prediction $\mathbf{h}(\mathbf{x})$ and compare this prediction and the actual observation in the likelihood, which also propagates this difference back to the beginning of the time window to perform an update. This formulation is the basis for deriving the so-called strong-constraint 4DVar schemes widely used for weather-prediction applications and the iterative ensemble smoothers used to history-match reservoir models in petroleum-engineering applications. Fig. 2.4 illustrates this alternative smoother formulation used by the ensemble version of 4DVar, En4DVar, and EnRML methods discussed below. This formulation also allows for measurements distributed over the assimilation window. In Sect. 2.5, we will see that this problem, and the ones in the following two sections, requires a modified form of Bayes' theorem.

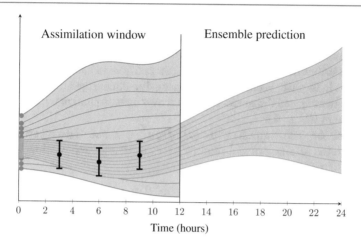

Fig. 2.4 The figure illustrates the recursive smoother formulation assuming a perfect model solved using an ensemble approach. One defines an assimilation time window and updates the initial conditions of the ensemble realizations at the beginning of the time window. Both the iterative ensemble smoothers and strong-constraint 4DVar use this approach. We based this graphic on an illustration from the ECMWF Forecast User Guide https://confluence.ecmwf.int/display/FUG/Forecast+User+Guide

2.4.5 Parameter Estimation

An analog problem to the smoother problem with a perfect model is the parameter-estimation problem for a model without uncertain controls and model errors. In this case, we define $\mathbf{z} = \boldsymbol{\theta}$ to contain the uncertain model parameters. This formulation leads to a situation where $\mathbf{z} = \boldsymbol{\theta}$, and we write, as usual,

$$\mathbf{y} = \mathbf{g}(\mathbf{z}) = \mathbf{h}\big(\mathbf{m}(\boldsymbol{\theta})\big), \tag{2.33}$$

where $\mathbf{m}(\boldsymbol{\theta})$ shows that we need the forward model to evaluate how the parameters relate to the measurements. For the practical implementation and solution, this problem is analogous to the "smoother problem for perfect models" solved for each assimilation window in Sect. 2.4.4. This problem formulation is also the one we solve in petroleum applications (Evensen et al., 2019; Evensen, 2021) and Chap. 21. Thus, Fig. 2.4, used to illustrate state estimation, also represents the parameter-estimation problem. For the Bayesian formulation of the parameter estimation problem, we refer to Sect. 2.5.

2.4.6 Estimating Initial Conditions, Parameters, Controls, and Errors

In the final formulation, we present the problem of estimating the initial condition for the time window, together with model parameters, model controls, and model errors. Evensen (2019) explained how we could solve the recursive smoother problem for a time window when we include model errors, while Evensen (2021) considered the case with additional control parameters. The controls can represent, for example, the imposed production rates or the aquifer strength in a reservoir model (Glegola et al., 2012; Peters et al., 2010) or the atmospheric forcing in ocean models (Vossepoel et al., 2004). Furthermore, Evensen (2021) illustrated how to include the model errors and controls in the state vector and simultaneously estimate the initial conditions $\mathbf{x}_0$, the model errors $\mathbf{q}$, and the model controls $\mathbf{u}$ over an assimilation window (see also the example in Chap. 21).

We define the state vector as $\mathbf{z}^T = \left(\mathbf{x}_0^T, \boldsymbol{\theta}^T, \mathbf{u}^T, \mathbf{q}^T\right)$, and given an estimate of $\mathbf{z}$, we compute the resulting model solution over the assimilation window, and hence the predicted measurements

$$\mathbf{y} = \mathbf{g}(\mathbf{z}) = \mathbf{h}\big(\mathbf{m}(\mathbf{x}_0, \boldsymbol{\theta}, \mathbf{u}, \mathbf{q})\big). \tag{2.34}$$

Again the practical implementation and solution of this problem, although more complicated due to the extended state vector, is analogous to the problems solved in the two previous Sects. 2.4.4 and 2.4.5, and we again refer to the modified form of Bayes' theorem in the following section.

2.5   Including the Predicted Measurements in Bayes Theorem

In Sects. 2.4.4, 2.4.5, and 2.4.6 we solve a problem involving a relation between the state vector $\mathbf{z}$ and the predicted measurements given by

$$\mathbf{y} = \mathbf{g}(\mathbf{z}) = \mathbf{h}\big(\mathbf{m}(\mathbf{z})\big). \tag{2.35}$$

In this case, it is not entirely clear how to evaluate the likelihood $f(\mathbf{d}|\mathbf{z})$ in Bayes' formula in Eq. (2.23). The reason is that we have measurements of the model state $\mathbf{x} = \mathbf{m}(\mathbf{z})$, which does not appear explicitly in Eq. (2.23). To solve this issue, we first write the likelihood to include the model state $\mathbf{x}$ explicitly as follows

$$f(\mathbf{d}|\mathbf{z}) = \int f(\mathbf{d}, \mathbf{x}|\mathbf{z}) \, d\mathbf{x} = \int f(\mathbf{d}|\mathbf{x}, \mathbf{z}) f(\mathbf{x}|\mathbf{z}) \, d\mathbf{x}, \tag{2.36}$$

where, if $\mathbf{z}$ contains the model initial condition $\mathbf{x}_0$, we exclude it from $\mathbf{x}$. For a given model state, the model inputs $\mathbf{z}$ do not provide extra information on the measurements. Thus, we can write

$$f(\mathbf{d}|\mathbf{x}, \mathbf{z}) = f(\mathbf{d}|\mathbf{x}), \tag{2.37}$$

leading to Bayes' theorem in the form

$$f(\mathbf{z}|\mathbf{d}) = \frac{f(\mathbf{z})}{f(\mathbf{d})} \int f(\mathbf{d}|\mathbf{x}) f(\mathbf{x}|\mathbf{z}) \, d\mathbf{x}. \tag{2.38}$$

This problem is costly to compute as we have to integrate the model state over all possible model inputs as dictated by $f(\mathbf{x}|\mathbf{z})$. For example, in the pure parameter estimation problem of Sect. 2.4.5, we would need to evaluate the integral over all possible model initial conditions and all possible model trajectories in the assimilation window. And we would have to do that for each parameter vector $\boldsymbol{\theta}$. To avoid this often intractable problem, one typically estimates both the model state $\mathbf{x}$ and inputs in $\mathbf{z}$.

It is sometimes helpful to rewrite the likelihood of variables in observation space instead of in state space. Recall that the likelihood $f(\mathbf{d}|\mathbf{z})$ is a function of $\mathbf{z}$ since we know the measurements $\mathbf{d}$ in the assimilation process. We can transform the variable $\mathbf{z}$ to observation space via the predicted measurements. Hence, we can augment the likelihood with the predicted measurements via

$$f(\mathbf{d}|\mathbf{z}) = \int f(\mathbf{d}, \mathbf{y}|\mathbf{z}) \, d\mathbf{y} = \int f(\mathbf{d}|\mathbf{y}, \mathbf{z}) f(\mathbf{y}|\mathbf{z}) \, d\mathbf{y}, \qquad (2.39)$$

where the second equality follows from the definition of a conditional density.

We first note that once we know the predicted measurements, the state vector $\mathbf{z}$ contains no new information on them, and we can write the likelihood

$$f(\mathbf{d}|\mathbf{y}, \mathbf{z}) = f(\mathbf{d}|\mathbf{y}). \qquad (2.40)$$

Furthermore, using the relation between the predicted measurements and the state vector, $\mathbf{y} = \mathbf{g}(\mathbf{z})$, we can write

$$f(\mathbf{y}|\mathbf{z}) = \delta(\mathbf{y} - \mathbf{g}(\mathbf{z})), \qquad (2.41)$$

where δ is the Dirac-delta function. We then have that a given state vector $\mathbf{z} = (\mathbf{x}_0, \boldsymbol{\theta}, \mathbf{u}, \mathbf{q})$ uniquely defines a set of predicted measurements $\mathbf{y}$. This equation is valid for predicted measurements over a whole assimilation window, also when we include model errors and controls as part of the state vector $\mathbf{z}$. Thus, it was a neat trick in Sect. 2.4.6 to include the model errors and controls in the state vector as a set of poorly known parameters (Evensen, 2019).

We now have the likelihood

$$f(\mathbf{d}|\mathbf{z}) = \int f(\mathbf{d}|\mathbf{y}) \delta(\mathbf{y} - \mathbf{g}(\mathbf{z})) \, d\mathbf{y} = f(\mathbf{d}|\mathbf{g}(\mathbf{z})), \qquad (2.42)$$

allowing us to write Bayes theorem as

Bayes' theorem related to the predicted measurements

$$f(\mathbf{z}|\mathbf{d}) = \frac{f(\mathbf{d}|\mathbf{g}(\mathbf{z})) \, f(\mathbf{z})}{f(\mathbf{d})}. \qquad (2.43)$$

To summarize, $\mathbf{z}$ is the state vector we try to estimate in the data-assimilation problem. In the general-smoother in Sect. 2.4.1, $\mathbf{z} = \mathbf{x}$ is the model state trajectory over the assimilation window. In contrast, in the filter formulation in Sect. 2.4.2, the state vector represents the model state at the end of the assimilation window $\mathbf{z} = \mathbf{x}_K$. For the recursive smoother in Sect. 2.4.3, $\mathbf{z}$ is the whole model state over the current and previous assimilation windows. Then in the smoother formulation for the perfect model in Sect. 2.4.4, where we estimate the initial state, the state vector is just the model initial conditions for the assimilation window $\mathbf{z} = \mathbf{x}_0$, which we consider as an input to the model. The situation is analogous in the parameter estimation problem in Sect. 2.4.5, where $\mathbf{z} = \boldsymbol{\theta}$, and in the final example in Sect. 2.4.6, where the state vector consists of all the uncertain model inputs $\mathbf{z}^T = (\mathbf{x}_0^T, \boldsymbol{\theta}^T, \mathbf{q}^T, \mathbf{u}^T)$.

The Bayes' formula in Eq. (2.43) applies to all these cases. For the state estimation examples in Sects. 2.4.1–2.4.3 we define $\mathbf{g}(\mathbf{z})$ as the measurement operator, while in the three last cases in Sects. 2.4.4–2.4.6, $\mathbf{g}(\mathbf{z})$ also includes the model integration. In the following, we will discuss popular data-assimilation methods that solve the Bayesian estimation problem in Eq. (2.43) under various approximations.

Maximum a Posteriori Solution

3

We will now introduce a fundamental approximation used in most practical data-assimilation methods, namely the definition of Gaussian priors. This approximation simplifies the Bayesian posterior, which allows us to compute the maximum a posteriori (MAP) estimate and sample from the posterior pdf. This chapter will introduce the Gaussian approximation and then discuss the Gauss–Newton method for finding the MAP estimate. This method is the starting point for many of the data-assimilation algorithms discussed in the following chapters.

3.1 Maximum a Posteriori (MAP) Estimate

The MAP solution is the state vector $\mathbf{z}$ that maximizes the posterior pdf, and can thus be seen as the most probable solution for $\mathbf{z}$ given the measurements $\mathbf{d}$. We define it as

$$\mathbf{z}_{\text{MAP}} = \underset{\mathbf{z}}{\text{argmax}}\left(f(\mathbf{z}|\mathbf{d})\right). \tag{3.1}$$

The variable $\mathbf{z}$ is called the *control variable* or *control vector* in the inverse modeling and control literature. Since we can write any smooth posterior pdf as

$$f(\mathbf{z}|\mathbf{d}) \propto \exp\left\{-\mathcal{J}(\mathbf{z})\right\}, \tag{3.2}$$

and the logarithm is a monotonically increasing function of its argument, the vector that maximizes the posterior pdf equals the vector that minimizes the *cost function* $\mathcal{J}(\mathbf{z})$. Hence, we can write

$$\mathbf{z}_{\text{MAP}} = \underset{\mathbf{z}}{\text{argmin}}\,\mathcal{J}(\mathbf{z}). \tag{3.3}$$

We can find a function's minimum by setting its gradient equal to zero. So, at the minimum, we have

$$\nabla_z \mathcal{J}(\mathbf{z}_{\text{MAP}}) = 0. \tag{3.4}$$

© The Author(s) 2022
G. Evensen et al., *Data Assimilation Fundamentals*,
Springer Textbooks in Earth Sciences, Geography and Environment,
https://doi.org/10.1007/978-3-030-96709-3_3

Furthermore, the second derivative of the cost function, the so-called *Hessian*, has information on the cost function's curvature at the minimum. As we will see, the inverse of that Hessian provides a first-order estimate of the posterior covariance.

In most geoscience applications of data assimilation that compute the MAP estimate, one assumes that both the prior and observation errors are Gaussian, leading to a more tractable problem. We will explore such methods in the following sections, followed by the explicit solutions for linear problems and an extensive treatment of iterative methods for nonlinear problems.

3.2 Gaussian Prior and Likelihood

Many popular data-assimilation methods assume that the prior distributions are Gaussian, leading to a simple representation of the data-assimilation problem. Note that the cost function is not quadratic in $\mathbf{z}$ as the measurement operator is still nonlinear. Hence, we introduce the following approximation.

Approximation 4 (Gaussian prior and likelihood) *We assume that the prior distributions of the state vector's components $\mathbf{z}$ and observation errors ϵ are both Gaussian distributed.*   □

We will in Chap. 9 discuss methods that do not apply Approx. 4. Now, we define

$$f(\mathbf{z}) = \mathcal{N}(\mathbf{z}^{\mathrm{f}}, \mathbf{C}_{zz}), \tag{3.5}$$

$$f(\mathbf{d} \mid \mathbf{g}(\mathbf{z})) = f(\epsilon) = \mathcal{N}(\mathbf{0}, \mathbf{C}_{dd}), \tag{3.6}$$

where the superscript f denote "first guess." Thus, $\mathbf{z}^{\mathrm{f}}$ is the "first guess" or *prior* estimate of the state vector, and $\mathbf{C}_{zz}$ is its error covariance. The prior error covariance includes the covariances between all the uncertain variables in the state vector,

$$\mathbf{C}_{zz} = \begin{pmatrix} \mathbf{C}_{x_0 x_0} & \mathbf{C}_{x_0 \theta} & \mathbf{C}_{x_0 u} & \mathbf{C}_{x_0 q} \\ \mathbf{C}_{\theta x_0} & \mathbf{C}_{\theta \theta} & \mathbf{C}_{\theta u} & \mathbf{C}_{\theta q} \\ \mathbf{C}_{u x_0} & \mathbf{C}_{u \theta} & \mathbf{C}_{u u} & \mathbf{C}_{u q} \\ \mathbf{C}_{q x_0} & \mathbf{C}_{q \theta} & \mathbf{C}_{q u} & \mathbf{C}_{q q} \end{pmatrix}. \tag{3.7}$$

Note that we formulate $\mathbf{z}$ such that it contains the model state at time zero and the model errors at other times, the so-called *forcing formulation*. In this case, the Gaussian prior assumption is reasonable and often assumed. However, if we reformulate the problem so that $\mathbf{z}$ contains the model solution, there is no model error in $\mathbf{z}$. A Gaussian prior for $\mathbf{z}$ would then force us to assume that the model is linear since only a linear model initialized with a Gaussian initial state would yield a model state that remains Gaussian over a time window. For this reason, we use the *forcing formulation* here. In most data-assimilation problems, we would neglect the covariances between different variables and retain only the covariance matrices on the diagonal. However, for the derivation of the methods below, we do not need to make this assumption.

The introduction of Gaussian priors leads to a posterior pdf formulation that we use in Bayes theorem to find

$$f(\mathbf{z}|\mathbf{d}) \propto \exp\left\{-\mathcal{J}(\mathbf{z})\right\}, \tag{3.8}$$

with the cost function $\mathcal{J}(\mathbf{z})$ defined as

Cost function

$$\mathcal{J}(\mathbf{z}) = \frac{1}{2}\left(\mathbf{z} - \mathbf{z}^{\mathrm{f}}\right)^{\mathrm{T}} \mathbf{C}_{zz}^{-1}\left(\mathbf{z} - \mathbf{z}^{\mathrm{f}}\right) + \frac{1}{2}\left(\mathbf{g}(\mathbf{z}) - \mathbf{d}\right)^{\mathrm{T}} \mathbf{C}_{dd}^{-1}\left(\mathbf{g}(\mathbf{z}) - \mathbf{d}\right). \tag{3.9}$$

Note that $\mathbf{g}(\mathbf{z})$ is the nonlinear mapping from the state vector, i.e., initial conditions, model errors, and parameters, to the predicted measurements. Thus, we have used the Bayesian formulation from Eq. (2.43). As mentioned, minimizing $\mathcal{J}(\mathbf{z})$ in Eq. (3.9) is equivalent to maximizing the *a posteriori* probability (MAP) solution of the posterior pdf in Eq. (3.8) with Approx. 4 on the Gaussian priors.

To find the MAP solution, we start with the cost function's gradient

$$\nabla_{\mathbf{z}}\mathcal{J}(\mathbf{z}) = \mathbf{C}_{zz}^{-1}\left(\mathbf{z} - \mathbf{z}^{\mathrm{f}}\right) + \nabla_{\mathbf{z}}\mathbf{g}(\mathbf{z})\,\mathbf{C}_{dd}^{-1}\left(\mathbf{g}(\mathbf{z}) - \mathbf{d}\right), \tag{3.10}$$

and by setting it to zero we define the minimizing solution $\mathbf{z}^{\mathrm{a}}$ from

---

The gradient set to zero

$$\mathbf{C}_{zz}^{-1}\left(\mathbf{z}^{\mathrm{a}} - \mathbf{z}^{\mathrm{f}}\right) + \nabla_{\mathbf{z}}\mathbf{g}(\mathbf{z}^{\mathrm{a}})\,\mathbf{C}_{dd}^{-1}\left(\mathbf{g}(\mathbf{z}^{\mathrm{a}}) - \mathbf{d}\right) = 0. \tag{3.11}$$

Here the superscript a denote "analysis." This equation forms the implicit, closed-form solution of our estimation problem that minimizes the cost function in Eq. (3.9). The model sensitivity $\nabla_{\mathbf{z}}\mathbf{g}(\mathbf{z}^{\mathrm{a}})$ is the gradient of the predicted measurements to the state vector. The following sections will present iterative methods that solve Eq. (3.11), leading to various 4DVar formulations.

3.3 Iterative Solutions

Even if $\mathbf{g}$ is a linear function of its argument and we can write down the explicit solution to Eq. (3.11), it is not uncommon to solve the problem iteratively. The reason is that the matrices involved can be of very high dimension and impossible to store in a computer. Another more practical reason is to avoid inverting matrices.

An important iterative minimization method is the so-called *Newton method*, which we can derive from a second-order Taylor expansion of the cost function. If we have an estimate of the minimum $\mathbf{z}^i$, we can improve this estimate by minimizing the expression

$$\mathcal{J}\left(\mathbf{z}^i + \delta\mathbf{z}\right) \approx \mathcal{J}\left(\mathbf{z}^i\right) + \delta\mathbf{z}^{\mathrm{T}}\,\nabla_{\mathbf{z}}\mathcal{J}^i + \frac{1}{2}\delta\mathbf{z}^{\mathrm{T}}\,\nabla_{\mathbf{z}}\nabla_{\mathbf{z}}\mathcal{J}^i\,\delta\mathbf{z}, \tag{3.12}$$

for $\delta \mathbf{z}$. Here $\nabla_{\mathbf{z}} \mathcal{J}^i$ denotes the cost function's gradient evaluated at $\mathbf{z}^i$, and $\nabla_{\mathbf{z}} \nabla_{\mathbf{z}} \mathcal{J}^i$ is the cost function's *Hessian* where we use the gradient operator twice. We readily find the solution for $\delta \mathbf{z}$ as

$$\frac{1}{2} \nabla_{\mathbf{z}} \nabla_{\mathbf{z}} \mathcal{J}^i \delta \mathbf{z} = -\nabla_{\mathbf{z}} \mathcal{J}^i. \tag{3.13}$$

The solution to this problem leads to a new estimate of the cost function's minimum, $\mathbf{z}^{i+1} = \mathbf{z}^i + \delta \mathbf{z}$, and we repeat the process with a second-order Taylor expansion around $\mathbf{z}^{i+1}$.

As mentioned above, although we could multiply this equation by the inverse of the Hessian to directly find the solution for $\delta \mathbf{z}$, this is often not the way this equation is solved. The reason is that the covariance matrices are often so large that they cannot be stored, not even on the world's most giant supercomputers. Instead, we use operators that return the matrix-vector products with these matrices. A beautiful example of this procedure is the so-called variational methods such as 4DVar, which replaces matrix-vector products with adjoint and forward model integrations. This procedure effectively leads to an iterative solution of the form

$$\mathbf{z}^{i+1} = \mathbf{z}^i - \gamma^i \mathbf{B}^i \nabla_{\mathbf{z}} \mathcal{J}\left(\mathbf{z}^i\right). \tag{3.14}$$

In Eq. (3.14), i is the iteration index, γ^i is a scalar that determines the so-called step size, and $\mathbf{B}^i$ is a matrix we can choose, typically in operator form, as mentioned above. The simplest choice, $\mathbf{B}^i = \mathbf{I}$, leads to the so-called *steepest descent* method, where the new iterate is directly downhill of the previous iterate. In many geoscience applications of data assimilation, this approach turns out to be a poor choice with a low convergence rate because the cost function often has a very irregular shape in high-dimensional spaces.

As we have seen, in the Newton method, one would like to choose $\mathbf{B}^i$ as the inverse of the Hessian, and $\gamma^i = 1$. The advantage of this choice is that if the Hessian is positive definite, the convergence rate is quadratic, meaning that $|\mathbf{z}^{i+1} - \mathbf{z}^{\mathrm{a}}| = r|\mathbf{z}^i - \mathbf{z}^{\mathrm{a}}|^2$, where $\mathbf{z}^{\mathrm{a}}$ denotes the state that minimizes the cost function and r is a positive constant that depends on details of the Hessian.

Often the Hessian is not available, so it is common to use an approximate Hessian. For instance, the *Gauss–Newton method* discussed below ignores part of the Hessian to ensure that the matrix in front of $\delta \mathbf{z}$ is symmetric positive definite by construction. Other approaches may start with an approximation to the Hessian and make this approximation more accurate at each iteration by using new gradient information. These are so-called *quasi-Newton methods*, and a much-used alternative is the Broyden, Fletcher, Goldfarb, and Shanno (BFGS) method. Because these methods use information from the Hessian, their convergence rate is faster than methods that ignore that information, such as steepest descent, but still not quadratic as in the Newton method. We say their convergence is superlinear.

If the matrix in front of $\delta \mathbf{z}$ is symmetric and positive definite, we can use an extremely efficient method called *conjugate gradient*. It has the advantage that it only requires the computation of one matrix-vector product at each iteration. Furthermore, we do not need to store the matrix. We can often represent it by a code that takes a vector as input and gives the matrix times that vector as output.

The Newton method is used in 3DVar and 4DVar, as we will discuss in Chaps. 4, 5, and 6. Primarily used, however, is the Gauss–Newton method, leading to implementations such as incremental 4DVar, which explores the conjugate-gradient minimization method commonly used in numerical weather and ocean forecasting. Furthermore, this formalism has led to a general methodology that can effectively be solved in ensemble space, resulting in iterative ensemble smoothers used in reservoir-engineering applications, amongst others. We will discuss this method next.

3.4 Gauss–Newton Iterations

A popular choice for finding an iterative solution to the cost function is the so-called Gauss–Newton method (Lawless et al., 2005). The Gauss–Newton method is an approximate Newton method where we approximate the Hessian by ignoring the second-order derivative of the nonlinear measurement operator. Let's take a deeper look at this approximation. We can write the full Hessian of the cost function in Eq. (3.9) as

$$\nabla_{\mathbf{z}} \nabla_{\mathbf{z}} \mathcal{J} = \mathbf{C}_{zz}^{-1} + \nabla_{\mathbf{z}} \mathbf{g}(\mathbf{z}) \, \mathbf{C}_{dd}^{-1} \left(\nabla_{\mathbf{z}} \mathbf{g}(\mathbf{z}) \right)^{\mathrm{T}} + \nabla_{\mathbf{z}} \nabla_{\mathbf{z}} \mathbf{g}(\mathbf{z}) \, \mathbf{C}_{dd}^{-1} \left(\mathbf{g}(\mathbf{z}) - \mathbf{d} \right). \quad (3.15)$$

The Gauss–Newton method ignores the last term, leading to

$$\nabla_{\mathbf{z}} \nabla_{\mathbf{z}} \mathcal{J}(\mathbf{z}) \approx \mathbf{C}_{zz}^{-1} + \nabla_{\mathbf{z}} \mathbf{g}(\mathbf{z}) \, \mathbf{C}_{dd}^{-1} \left(\nabla_{\mathbf{z}} \mathbf{g}(\mathbf{z}) \right)^{\mathrm{T}}. \quad (3.16)$$

We can now write a Gauss–Newton iteration similar to Eq. (3.14) as

---

Gauss–Newton iteration

$$\mathbf{z}^{i+1} = \mathbf{z}^{i} - \gamma^{i} \left(\mathbf{C}_{zz}^{-1} + \mathbf{G}^{i\,\mathrm{T}} \mathbf{C}_{dd}^{-1} \mathbf{G}^{i} \right)^{-1} \left(\mathbf{C}_{zz}^{-1} (\mathbf{z}^{i} - \mathbf{z}^{\mathrm{f}}) + \mathbf{G}^{i\,\mathrm{T}} \mathbf{C}_{dd}^{-1} \left(\mathbf{g}(\mathbf{z}^{i}) - \mathbf{d} \right) \right). \quad (3.17)$$

Here the increment is a steplength γ times the gradient normalized by $(\mathbf{C}_{zz}^{-1} + \mathbf{G}^{i\,\mathrm{T}} \mathbf{C}_{dd}^{-1} \mathbf{G}^{i})$, the approximate Hessian. In correspondence with Eq. (3.14), we have chosen $\mathbf{B}^{i} = \left(\mathbf{C}_{zz}^{-1} + \mathbf{G}^{i\,\mathrm{T}} \mathbf{C}_{dd}^{-1} \mathbf{G}^{i} \right)^{-1}$. Furthermore, we have defined the gradient of $\mathbf{g}(\mathbf{z})$ at iteration i as

$$\mathbf{G}^{i\,\mathrm{T}} = \nabla_{\mathbf{z}} \mathbf{g}(\mathbf{z}^{i}). \quad (3.18)$$

We can interpret the operator $\mathbf{G}^{i}$ as the tangent-linear-model operator at iteration i, which provides the linear relation between the state vector and the observations. Likewise, we can interpret the operator $\mathbf{G}^{i\,\mathrm{T}}$ as the tangent-linear model's adjoint.

3.5 Incremental Form of Gauss–Newton Iterations

As mentioned earlier, the storage of the approximate Hessian would require substantial memory if we use the direct Gauss–Newton method for high-dimensional problems, and the Hessian's inversion can be rather expensive. We will present two solutions to solve this problem. In Chap 7, we will use Eq. (3.17) to develop the ensemble-random-maximum-likelihood (EnRML) method, which is commonly used in the petroleum industry.

An alternative is to write Eq. (3.17) with $\gamma^i = 1$ as

$$\left(\mathbf{C}_{zz}^{-1} + \mathbf{G}^{i\mathrm{T}}\mathbf{C}_{dd}^{-1}\mathbf{G}^i\right)\left(\mathbf{z}^{i+1} - \mathbf{z}^i\right) = -\left(\mathbf{C}_{zz}^{-1}(\mathbf{z}^i - \mathbf{z}^f) + \mathbf{G}^{i\mathrm{T}}\mathbf{C}_{dd}^{-1}\left(\mathbf{g}(\mathbf{z}^i) - \mathbf{d}\right)\right). \quad (3.19)$$

When we define, as before,

$$\delta\mathbf{z} = \mathbf{z}^{i+1} - \mathbf{z}^i, \quad (3.20)$$

this equation also arises as the minimum of the following *quadratic* cost function for $\delta\mathbf{z}$

$$\begin{aligned}
\mathcal{J}(\delta\mathbf{z}) = &\frac{1}{2}\left(\delta\mathbf{z} + \mathbf{z}^i - \mathbf{z}^f\right)^{\mathrm{T}}\mathbf{C}_{zz}^{-1}\left(\delta\mathbf{z} + \mathbf{z}^i - \mathbf{z}^f\right) \\
&+ \frac{1}{2}\left(\mathbf{G}^i\delta\mathbf{z} + \mathbf{g}(\mathbf{z}^i) - \mathbf{d}\right)^{\mathrm{T}}\mathbf{C}_{dd}^{-1}\left(\mathbf{G}^i\delta\mathbf{z} + \mathbf{g}(\mathbf{z}^i) - \mathbf{d}\right).
\end{aligned} \quad (3.21)$$

This cost function linearizes the model and observation operators around the model trajectory for each Gauss–Newton iteration starting from the initial condition $\mathbf{z}^i$. Because $\delta\mathbf{z}$ is small, we can approximate $\mathbf{g}(\mathbf{z}^i + \delta\mathbf{z}) \approx \mathbf{g}(\mathbf{z}^i) + \mathbf{G}^i\delta\mathbf{z}$ in which $\mathbf{G}^i$ is the transpose of the gradient of $\mathbf{g}(\mathbf{z}^i)$ from Eq. (3.18). For convenience, we define the innovation vector

$$\boldsymbol{\eta}^i = \mathbf{d} - \mathbf{g}(\mathbf{z}^i), \quad (3.22)$$

and the residual

$$\boldsymbol{\xi}^i = \mathbf{z}^f - \mathbf{z}^i. \quad (3.23)$$

With $\boldsymbol{\eta}$ and $\boldsymbol{\xi}$, we can now write the cost function in Eq. (3.21) for the increments $\delta\mathbf{z}$ as

---

Quadratic cost function for the increments

$$\mathcal{J}(\delta\mathbf{z}) = \frac{1}{2}\left(\delta\mathbf{z} - \boldsymbol{\xi}^i\right)^{\mathrm{T}}\mathbf{C}_{zz}^{-1}\left(\delta\mathbf{z} - \boldsymbol{\xi}^i\right) + \frac{1}{2}\left(\mathbf{G}^i\delta\mathbf{z} - \boldsymbol{\eta}^i\right)^{\mathrm{T}}\mathbf{C}_{dd}^{-1}\left(\mathbf{G}^i\delta\mathbf{z} - \boldsymbol{\eta}^i\right). \quad (3.24)$$

The solution for the increments becomes, from Eq. (3.19),

$$\left(\mathbf{C}_{zz}^{-1} + \mathbf{G}^{i\mathrm{T}}\mathbf{C}_{dd}^{-1}\mathbf{G}^i\right)\delta\mathbf{z} = \mathbf{C}_{zz}^{-1}\boldsymbol{\xi}^i + \mathbf{G}^{i\mathrm{T}}\mathbf{C}_{dd}^{-1}\boldsymbol{\eta}^i. \quad (3.25)$$

We can solve this linear set of equations iteratively, and we usually implement the approximate Hessian as a set of operations working on the vector $\delta\mathbf{z}$. Quasi-Newton methods like BFGS and conjugate gradient are highly efficient for minimizing this cost function.

Thus, the incremental form of the Gauss–Newton method corresponds to an iterative scheme where we find the minimum of a quadratic cost function for $\delta\mathbf{z}$ in each iteration. After that, we update $\mathbf{z}^{i+1} = \mathbf{z}^i + \delta\mathbf{z}$ from (3.20), integrate the nonlinear model with the updated state vector, and recompute the variables $\boldsymbol{\eta}^i$ and $\boldsymbol{\xi}^i$ from Eqs. (3.22) and (3.23) before we solve the quadratic minimization problem again.

Gauss–Newton has a special status among minimization methods. It turns non-quadratic minimization problems into a sequence of quadratic minimization problems. We can solve each of these quadratic problems iteratively, leading to one iteration within another. We will explore this approach in the methods discussed in the following chapters.

Strong-Constraint 4DVar

4

This chapter introduces the *strong-constraint 4-dimensional variational* (SC-4DVar) method. By strong constraint, we refer to the dynamical model having no model errors. Hence, the model solution over the assimilation window is entirely determined by the model as soon as we give the initial conditions. In SC-4DVar, we solve a 4-dimensional problem by including the three space dimensions and time as the fourth dimension, using a variational approach. The method is a gradient-based minimization method but makes use of an adjoint model to calculate the gradient. The chapter covers the SC-4DVar's standard form for estimating initial conditions and uncertain parameters. After that, it discusses a more efficient incremental formulation before presenting the state-transform variant of the method.

4.1 Standard Strong-Constraint 4DVar Method

Sasaki (1970a) introduced the concept of a strong-constraint formulation for a minimization problem when imposing a dynamical model without errors as a strong constraint. The iterative SC-4DVar method for solving the strong-constraint problem has its origin in several publications in the atmosphere and ocean modeling communities, e.g., (Lewis & Derber, 1985; Le Dimet & Talagrand, 1986; Talagrand & Courtier, 1987; Thacker, 1988; Thacker & Long, 1988). Later, in Chap. 5, we will extend this formulation to the weak-constraint case where we allow the model to contain errors leading to the so-called *weak-constraint variational inverse problem* and the *weak-constraint 4DVar* (WC-4DVar) method.

© The Author(s) 2022
G. Evensen et al., *Data Assimilation Fundamentals*,
Springer Textbooks in Earth Sciences, Geography and Environment,
https://doi.org/10.1007/978-3-030-96709-3_4

4.1.1 Data-Assimilation Problem

We now assume the model system to include Eqs. (2.1, 2.2, and 2.5) and write it as

$$\mathbf{x}_0 = \mathbf{x}_0^f + \mathbf{x}_0', \tag{4.1}$$

$$\boldsymbol{\theta} = \boldsymbol{\theta}^f + \boldsymbol{\theta}', \tag{4.2}$$

$$\mathbf{x}_{k+1} = \mathbf{m}(\mathbf{x}_k, \boldsymbol{\theta}), \tag{4.3}$$

where there are no model errors or uncertainty in the model-state evolution, but we allow for uncertain model initial conditions and parameters. Additional constraints come from the measurements with errors

$$\mathbf{d} = \mathbf{h}(\mathbf{x}) + \mathbf{e}. \tag{4.4}$$

Thus, we wish to estimate the model's uncertain initial conditions at the start of the assimilation window and the poorly known parameters to find a model prediction close to the measurements. At the same time, the estimated initial conditions and parameters should remain close to their first-guess values while respecting the prescribed uncertainties in both. The state vector $\mathbf{z}$ contains the initial state and model parameters,

$$\mathbf{z} = \begin{pmatrix} \mathbf{x}_0 \\ \boldsymbol{\theta} \end{pmatrix}. \tag{4.5}$$

We start from the cost function in Eq. (3.9). The operator $\mathbf{g}(\mathbf{z})$ is the composite function including the model recursion from Eq. (4.3) that predicts the model solution at all time steps over the assimilation window followed by a measurement operator that maps the prediction to the measurements.

From the definition of the predicted measurements in Eq. (2.32), $\mathbf{g}(\mathbf{z}) = \mathbf{h}(\mathbf{m}(\mathbf{z}))$, we can write $\mathbf{g}(\mathbf{z}) = \mathbf{h}(\mathbf{x})$, and it is then convenient to reformulate the problem defined by the cost function in Eq. (3.9) as

---

SC-4DVar costfunction

$$\mathcal{J}(\mathbf{z}) = \frac{1}{2}\left(\mathbf{z} - \mathbf{z}^f\right)^{\mathrm{T}} \mathbf{C}_{zz}^{-1}\left(\mathbf{z} - \mathbf{z}^f\right) + \frac{1}{2}\left(\mathbf{h}(\mathbf{x}) - \mathbf{d}\right)^{\mathrm{T}} \mathbf{C}_{dd}^{-1}\left(\mathbf{h}(\mathbf{x}) - \mathbf{d}\right), \tag{4.6}$$

---

subject to the "perfect-model" constraint in Eq. (4.3), which defines the model solution $\mathbf{x}$ over the assimilation window.

In SC-4DVar, we commonly refer to the state-covariance matrix

$$\mathbf{C}_{zz} = \begin{pmatrix} \mathbf{C}_{x_0 x_0} & \mathbf{0} \\ \mathbf{0} & \mathbf{C}_{\theta\theta} \end{pmatrix}, \tag{4.7}$$

as the *background-error-covariance matrix*, which characterizes the error covariances of the prior initial conditions and the parameters. We specify the background-error-covariance matrix using time-independent numerical representations of prescribed relationships between variables (Weaver et al., 2003, 2005). We would typically not assume correlations between model parameters and the initial conditions.

4.1.2 Lagrangian Formulation

Minimizing the cost function in Eq. (4.6), subject to the additional constraint of the model Eq. (4.3), allows us to formulate a Lagrangian minimization problem with the model constraint introduced via Lagrangian multipliers $\boldsymbol{\lambda}_k$. Note that we already have included the prior initial condition and parameters in the first term of the cost function. By introducing the Lagrangian multipliers, we increase the number of unknowns in the optimization problem, but the formulation allows for an efficient solution method. The Lagrangian cost function for the constrained minimization problem becomes

$$
\begin{aligned}
\mathcal{L}(\mathbf{x}_0, \ldots, \mathbf{x}_{K+1}, \boldsymbol{\theta}, \boldsymbol{\lambda}_1, \ldots, \boldsymbol{\lambda}_{K+1}) =\ & \frac{1}{2}\left(\mathbf{x}_0 - \mathbf{x}_0^{\mathrm{f}}\right)^{\mathrm{T}} \mathbf{C}_{x_0 x_0}^{-1}\left(\mathbf{x}_0 - \mathbf{x}_0^{\mathrm{f}}\right) \\
& + \frac{1}{2}\left(\boldsymbol{\theta} - \boldsymbol{\theta}^{\mathrm{f}}\right)^{\mathrm{T}} \mathbf{C}_{\theta\theta}^{-1}\left(\boldsymbol{\theta} - \boldsymbol{\theta}^{\mathrm{f}}\right) \\
& + \frac{1}{2}\left(\mathbf{h}(\mathbf{x}) - \mathbf{d}\right)^{\mathrm{T}} \mathbf{C}_{dd}^{-1}\left(\mathbf{h}(\mathbf{x}) - \mathbf{d}\right) \\
& + \sum_{k=0}^{K} \boldsymbol{\lambda}_{k+1}^{\mathrm{T}}\left(\mathbf{x}_{k+1} - \mathbf{m}(\mathbf{x}_k, \boldsymbol{\theta})\right).
\end{aligned}
\tag{4.8}
$$

The last expression in this Lagrangian introduces the Lagrange multipliers and the perfect-model constraints. In the summation, we include an extra time step for $\mathbf{x}_{K+1}$ and $\boldsymbol{\lambda}_{K+1}$, leading to a more straightforward form of the Euler–Lagrange equations below.

We now define the gradients of the nonlinear measurement operator $\mathbf{h}$ and model $\mathbf{m}$ as

$$
\mathbf{H} = \nabla_{\mathbf{x}} \mathbf{h}(\mathbf{x})\big|_{\mathbf{x}} \ ,
\tag{4.9}
$$

$$
\mathbf{M}_{x,k} = \nabla_{\mathbf{x}_k} \mathbf{m}(\mathbf{x}, \boldsymbol{\theta})\big|_{\mathbf{x}_k, \boldsymbol{\theta}} \ ,
\tag{4.10}
$$

$$
\mathbf{M}_{\theta,k} = \nabla_{\theta} \mathbf{m}(\mathbf{x}, \boldsymbol{\theta})\big|_{\mathbf{x}_k, \boldsymbol{\theta}} \ ,
\tag{4.11}
$$

and simplify the notation $(\mathbf{x}_0, \ldots, \mathbf{x}_{K+1}, \boldsymbol{\theta}, \boldsymbol{\lambda}_1, \ldots, \boldsymbol{\lambda}_{K+1})$ by the more compact notation $(\mathbf{x}, \boldsymbol{\theta}, \boldsymbol{\lambda})$. The time index on $\mathbf{M}_{\theta,k}$ denotes that we evaluate the gradient to $\boldsymbol{\theta}$ at different times t_k.

4.1.3 Explaining the Measurement Operator

We assume that $\mathbf{H}$ consists of matrices $\mathbf{H}_k$, each with size $m \times n$, with m being the number of measurements within an assimilation window, and n the size of the state vector at the time t_k. We then define one matrix $\mathbf{H}_k$ for each time step t_k, that relates the predicted measurements at t_k to the state vector at $\mathbf{x}_k$. Thus,

$$
\mathbf{H} = \left(\mathbf{H}_0 \ \cdots \ \mathbf{H}_k \ \cdots \ \mathbf{H}_{K+1} \right).
\tag{4.12}
$$

The rows in $\mathbf{H}$ correspond to the m measurements distributed over the assimilation window, and $\mathbf{H}_k$ corresponds to measurements available at the time step k within this window. Thus, for a time, t_k, we can have a set of measurements, and a sub-block $\mathbf{H}_k$ will relate these measurements to the model state at that time. If there are no measurements at a time t_k, then $\mathbf{H}_k = \mathbf{0}$. Because of this construction, each matrix $\mathbf{H}_k$ is very sparse. All rows are zero except for the rows corresponding to the measurements at time t_k and the measurement location.

The matrix $\mathbf{H}$ can take care of interpolation between the measurement location and the model discretization. In contrast, suppose the measurement is taken precisely at a model gridpoint. In that case, the corresponding row in $\mathbf{H}$ will have a single element that connects the model variable to the measurement. In this manner, $\mathbf{y} = \mathbf{Hx}$ denotes the vector of all predicted measurements over the assimilation window. Likewise, $\mathbf{y}_k = \mathbf{H}_k \mathbf{x}_k$ is the vector $\mathbf{y}$ with zeros except for the predicted measurements at time t_k at the measurement location. Note that we have defined $\mathbf{H}_0 = \mathbf{H}_{K+1} = \mathbf{0}$. We have used this definition for $\mathbf{H}$ to allow for a compact and straightforward notation in the following and at the same time allow for measurement errors correlated over time.

4.1.4 Euler–Lagrange Equations

We now find for the gradient of the Lagrangian to $\mathbf{x}_k$, for $k = 1, \ldots, K$,

$$\nabla_{\mathbf{x}_k} \mathcal{L}(\mathbf{x}, \boldsymbol{\theta}, \boldsymbol{\lambda}) = \mathbf{H}_k^{\mathrm{T}} \mathbf{C}_{dd}^{-1} \left(\mathbf{h}(\mathbf{x}) - \mathbf{d} \right) + \boldsymbol{\lambda}_k - \mathbf{M}_{x,k}^{\mathrm{T}} \boldsymbol{\lambda}_{k+1}. \qquad (4.13)$$

The transpose of the linearized model $\mathbf{M}_{x,k}$, known as the *adjoint* of the model, deserves some specific attention. The linearized model $\mathbf{M}_{x,k}$ maps a vector from time t_k to time t_{k+1}. Its adjoint, $\mathbf{M}_{x,k}^{\mathrm{T}}$, does the reverse, it maps a vector from time t_{k+1} backwards in time to t_k. Hence, Eq. (4.13) refers to a backward integration of the linearized model equations' adjoint model.

The gradient of the Lagrangian to $\mathbf{x}_{K+1}$ becomes simply

$$\nabla_{\mathbf{x}_{K+1}} \mathcal{L}(\mathbf{z}, \mathbf{x}, \boldsymbol{\lambda}) = \boldsymbol{\lambda}_{K+1}. \qquad (4.14)$$

For the initial time, we obtain the gradient of the cost function to $\mathbf{x}_0$ as

$$\begin{aligned}
\nabla_{\mathbf{x}_0} \mathcal{L}(\mathbf{x}, \boldsymbol{\theta}, \boldsymbol{\lambda}) &= \mathbf{C}_{zz}^{-1} \left(\mathbf{x}_0 - \mathbf{x}_0^{\mathrm{f}} \right) - \mathbf{M}_{x,0}^{\mathrm{T}} \boldsymbol{\lambda}_1 \\
&= \mathbf{C}_{zz}^{-1} \left(\mathbf{x}_0 - \mathbf{x}_0^{\mathrm{f}} \right) - \boldsymbol{\lambda}_0,
\end{aligned} \qquad (4.15)$$

where we have defined an additional "pseudo-variable" $\boldsymbol{\lambda}_0$ for convenience. The derivative of the Lagrangian to the parameters $\boldsymbol{\theta}$ gives

$$\nabla_{\boldsymbol{\theta}} \mathcal{L}(\mathbf{x}, \boldsymbol{\theta}, \boldsymbol{\lambda}) = \mathbf{C}_{\theta\theta}^{-1} \left(\boldsymbol{\theta} - \boldsymbol{\theta}^{\mathrm{f}} \right) - \sum_{k=0}^{K} \mathbf{M}_{\theta,k}^{\mathrm{T}} \boldsymbol{\lambda}_{k+1}. \qquad (4.16)$$

Finally, the derivative of $\mathcal{L}(\mathbf{x}, \boldsymbol{\theta}, \boldsymbol{\lambda})$ to the Lagrange multiplier $\boldsymbol{\lambda}_k$ returns the model equation

$$\nabla_{\boldsymbol{\lambda}_k} \mathcal{L}(\mathbf{x}, \boldsymbol{\theta}, \boldsymbol{\lambda}) = \mathbf{x}_{k+1} - \mathbf{m}(\mathbf{x}_k, \boldsymbol{\theta}). \qquad (4.17)$$

Algorithm 1 Standard SC-4DVar algorithm with parameter estimation

1: Input: $\mathbf{z}^{\mathrm{f}} \in \mathfrak{R}^n$; $\mathbf{d} \in \mathfrak{R}^m$ ▷ Prior initial conditions and observations
2: $\mathbf{x}_0 = \mathbf{x}_0^{\mathrm{f}}$ ▷ Initialization of $\mathbf{x}_0$
3: $\boldsymbol{\theta} = \boldsymbol{\theta}^{\mathrm{f}}$ ▷ Initialization of $\boldsymbol{\theta}$
4: **repeat**   ▷ Iteration loop
5:      **for** $k = 0 : K$ **do**   ▷ Integrate forward model
6:          $\mathbf{x}_{k+1} = \mathbf{m}(\mathbf{x}_k, \boldsymbol{\theta})$
7:      **end for**
8:      $\lambda_{K+1} = 0$
9:      **for** $k = K : 0$ **do**   ▷ Integrate backward adjoint model
10:         $\lambda_k = \mathbf{M}_{x,k}^{\mathrm{T}} \lambda_{k+1} - \mathbf{H}_k^{\mathrm{T}} \mathbf{C}_{dd}^{-1} (\mathbf{h}(\mathbf{x}) - \mathbf{d})$
11:     **end for**
12:     $\mathbf{x}_0 \leftarrow \mathbf{x}_0 - \gamma \mathbf{B} \nabla_{x_0} \mathcal{L}(\mathbf{x}, \boldsymbol{\theta}, \lambda)$ ▷ Update $\mathbf{x}_0$ using Eq. (4.15)
13:     $\boldsymbol{\theta} \leftarrow \boldsymbol{\theta} - \gamma \mathbf{B} \nabla_{\theta} \mathcal{L}(\mathbf{x}, \boldsymbol{\theta}, \lambda)$ ▷ Update $\boldsymbol{\theta}$ using Eq. (4.16)
14: **until** convergence

Setting the derivatives in Eqs. (4.13–4.17) to zero results in a coupled system of Euler–Lagrange equations consisting of a forward model

$$\mathbf{x}_0 = \mathbf{x}_0^{\mathrm{f}} + \mathbf{C}_{x_0 x_0} \lambda_0, \tag{4.18}$$

$$\boldsymbol{\theta} = \boldsymbol{\theta}^{\mathrm{f}} + \mathbf{C}_{\theta\theta} \sum_{k=0}^{K} \mathbf{M}_{\theta,k}^{\mathrm{T}} \lambda_{k+1}, \tag{4.19}$$

$$\mathbf{x}_{k+1} = \mathbf{m}(\mathbf{x}_k, \boldsymbol{\theta}), \tag{4.20}$$

and a backward model for the adjoint variable

$$\lambda_{K+1} = 0, \tag{4.21}$$

$$\lambda_k = \mathbf{M}_{x,k}^{\mathrm{T}} \lambda_{k+1} - \mathbf{H}_k^{\mathrm{T}} \mathbf{C}_{dd}^{-1} (\mathbf{h}(\mathbf{x}) - \mathbf{d}). \tag{4.22}$$

In this manner, we must solve a coupled two-point boundary-value problem in time. The last term of the right-hand side of Eq. (4.22) is often referred to as the weighted observational forcing (Daley, 1991; Talagrand & Courtier, 1987) and introduces the observation information, which is brought backward from an observation time to the start of the time window.

In the standard form of SC-4DVar, we use the gradients in Eqs. (4.15) and (4.16) in a Gauss–Newton method to iteratively update the initial conditions of the forward model. Thus, starting from the first-guess solution of the model with $\lambda = 0$, we obtain a solution for $\mathbf{x}$ from the forward model Eqs. (4.18–4.20). We can then use this $\mathbf{x}$ to solve the adjoint model in Eqs. (4.21) and (4.22) for λ. When we have an estimate of λ, we can evaluate the gradients in Eqs. (4.15) and (4.16) to update the initial condition and estimate the parameters and repeat the procedure. We typically use a conjugate-gradient or a quasi-Newton method for this iterative procedure (Navon & Legler, 1987). The linearization points for the observation operator and the model, defined by the states at different times in the forward model integration, differ in each iteration, such that $\mathbf{M}_{x,k}^{\mathrm{T}}$ and $\mathbf{M}_{\theta,k}$, and in some cases, $\mathbf{H}$ will differ between iterations. In Algorithm 1, we illustrate the practical implementation of the standard SC-4DVar method.

4.2 Incremental Strong-Constraint 4DVar

In the cost function in Eq. (4.6), both the model and the measurement operators are often nonlinear. Minimizing a non-quadratic cost function can be challenging as the most efficient minimization methods, such as conjugate gradient, assume a quadratic cost function. Therefore, an approach based on the incremental Gauss–Newton formulation can lead to a more straightforward minimization problem and more efficient solvers. Thus, to practically implement SC-4DVar, one often uses the more efficient incremental form of the Gauss–Newton method from Sect. 3.5, leading to the so-called incremental 4DVar (see, e.g., Weaver et al. 2005).

4.2.1 Incremental Formulation

Incremental 4DVar is particularly suitable for estimating initial conditions. If we want to estimate model parameters using the incremental approach, we need to update them in the outer iterations. The inner iterations perform a linearized model integration for the increments utilizing the model's tangent-linear operator evaluated at the current model solution $\mathbf{x}^i$ and parameters $\boldsymbol{\theta}^i$. Before addressing the parameter-estimation problem, let us focus on estimating the initial model state only. In this case, the state vector is $\mathbf{z} = \mathbf{x}_0$, and the dynamical model with an uncertain initial condition is now

$$\mathbf{x}_0 = \mathbf{x}_0^{\mathrm{f}} + \mathbf{x}_0', \tag{4.23}$$

$$\mathbf{x}_{k+1} = \mathbf{m}(\mathbf{x}_k). \tag{4.24}$$

In incremental SC-4DVar, we compute updates

$$\mathbf{z}^{i+1} = \mathbf{z}^i + \delta\mathbf{z}, \tag{4.25}$$

where the increments $\delta\mathbf{z}$ are solutions that minimize the cost function in Eq. (3.24) for iteration i.

As in Sect. 4.1, $\mathbf{g}(\mathbf{z})$ is the composite function of the recursive time stepping of the model (see Eq. 2.32). Thus, we find the model solution $\mathbf{x}$ over the assimilation window from Eq. (4.24) and apply the measurement operator $\mathbf{h}(\mathbf{x})$ to obtain the predicted measurements. The linearization of $\mathbf{m}$ now gives the tangent-linear operator of the nonlinear model evaluated at the model solution $\mathbf{x}^i$, from the ith outer iteration

$$\mathbf{M}_k^i = \nabla_{\mathbf{x}_k} \mathbf{m}(\mathbf{x})\Big|_{\mathbf{x}_k^i}, \tag{4.26}$$

similar to the definition in Eq. (4.10).

We will now minimize the cost function in Eq. (3.24) iteratively. We can compute the model solution's ith increment $\delta\mathbf{x} = \delta\mathbf{x}_1, ..., \delta\mathbf{x}_{K+1}$ over the assimilation window from the tangent-linear model with initial conditions

$$\delta\mathbf{x}_0 = \delta\mathbf{z}^i, \tag{4.27}$$

$$\delta\mathbf{x}_{k+1} = \mathbf{M}_k^i \delta\mathbf{x}_k. \tag{4.28}$$

For the predicted measurements of the increments, we can write $\mathbf{H}\delta\mathbf{x}$, using the linearized measurement operator $\mathbf{H}$ from Eq. (4.9). Additionally, for the ith iteration, we define the prior increment

$$\boldsymbol{\xi}^i = \mathbf{x}_0^{\mathrm{f}} - \mathbf{x}_0^i, \tag{4.29}$$

and the innovation

$$\boldsymbol{\eta}^i = \mathbf{d} - \mathbf{h}(\mathbf{x}^i). \tag{4.30}$$

For each iteration i, the problem reduces to minimizing the cost function

---

Inner incremental SC-4DVar costfunction

$$\mathcal{J}(\delta\mathbf{z}) = \frac{1}{2}\left(\delta\mathbf{z} - \boldsymbol{\xi}^i\right)^{\mathrm{T}} \mathbf{C}_{zz}^{-1}\left(\delta\mathbf{z} - \boldsymbol{\xi}^i\right) + \frac{1}{2}\left(\mathbf{H}^i \delta\mathbf{x} - \boldsymbol{\eta}^i\right)^{\mathrm{T}} \mathbf{C}_{dd}^{-1}\left(\mathbf{H}^i \delta\mathbf{x} - \boldsymbol{\eta}^i\right), \tag{4.31}$$

---

for $\delta\mathbf{z}$, subject to the model constraint in Eq. (4.28). Note the similarity to Eq. (3.24).

4.2.2 Lagrangian Formulation for the Inner Iterations

We again introduce Lagrange multipliers to form the extended cost function that incorporates the strong constraint of the perfect model, similar to Eq. (4.8). We also define the additional control variables $\delta\boldsymbol{\lambda} = \delta\boldsymbol{\lambda}_1, ..., \delta\boldsymbol{\lambda}_{K+1}$. The Lagrangian cost function for the problem defined by Eqs. (4.28) and (4.31) now becomes

$$\begin{aligned} \mathcal{L}(\delta\mathbf{z}, \delta\mathbf{x}, \delta\boldsymbol{\lambda}) = {} & \frac{1}{2}\left(\delta\mathbf{z} - \boldsymbol{\xi}^i\right)^{\mathrm{T}} \mathbf{C}_{zz}^{-1}\left(\delta\mathbf{z} - \boldsymbol{\xi}^i\right) \\ & + \frac{1}{2}\left(\mathbf{H}^i \delta\mathbf{x} - \boldsymbol{\eta}^i\right)^{\mathrm{T}} \mathbf{C}_{dd}^{-1}\left(\mathbf{H}^i \delta\mathbf{x} - \boldsymbol{\eta}^i\right) \\ & + \sum_{k=0}^{K} \delta\boldsymbol{\lambda}_{k+1}^{\mathrm{T}}\left(\delta\mathbf{x}_{k+1} - \mathbf{M}_k^i \delta\mathbf{x}_k\right). \end{aligned} \tag{4.32}$$

The gradient of the Lagrangian for the incremental problem at time k becomes

$$\nabla_{\delta\mathbf{x}_k}\mathcal{L}(\delta\mathbf{z}, \delta\mathbf{x}, \delta\boldsymbol{\lambda}) = \mathbf{H}_k^{i\,\mathrm{T}} \mathbf{C}_{dd}^{-1}\left(\mathbf{H}^i \delta\mathbf{x} - \boldsymbol{\eta}^i\right) + \delta\boldsymbol{\lambda}_k - \mathbf{M}_k^{i\,\mathrm{T}} \delta\boldsymbol{\lambda}_{k+1}. \tag{4.33}$$

Similarly, we have for the final time t_{K+1},

$$\nabla_{\delta\mathbf{x}_K}\mathcal{L}(\delta\mathbf{z}, \delta\mathbf{x}, \delta\boldsymbol{\lambda}) = \delta\boldsymbol{\lambda}_{K+1}, \tag{4.34}$$

and for t_0 we find

$$\nabla_{\delta \mathbf{z}} \mathcal{L}(\delta \mathbf{z}, \delta \mathbf{x}, \delta \boldsymbol{\lambda}) = \mathbf{C}_{zz}^{-1}\left(\delta \mathbf{z} - \boldsymbol{\xi}^i\right) - \mathbf{M}_0^{i\,\mathrm{T}} \delta \boldsymbol{\lambda}_1$$
$$= \mathbf{C}_{zz}^{-1}\left(\delta \mathbf{z} - \boldsymbol{\xi}^i\right) - \delta \boldsymbol{\lambda}_0, \tag{4.35}$$

using $\delta \mathbf{z} = \delta \mathbf{x}_0$. Finally, the derivatives of the Lagrangian to the Lagrange multipliers give the linearized forward model Eq. (4.28) in a similar way as how we arrived at Eq. (4.17) for the non-incremental 4DVar formulation.

4.2.3 Euler–Lagrange Equations for the Inner Iterations

Setting the derivatives in Eqs. (4.33), (4.34), and (4.35), and also the derivatives of the Lagrangian to $\delta \boldsymbol{\lambda}_k$, all equal to zero, gives the following set of coupled Euler–Lagrange equations consisting of the linear forward model

$$\delta \mathbf{x}_0 = \boldsymbol{\xi}^i + \mathbf{C}_{zz}\, \delta \boldsymbol{\lambda}_0, \tag{4.36}$$

$$\delta \mathbf{x}_{k+1} - \mathbf{M}_k^i \delta \mathbf{x}_k = 0, \tag{4.37}$$

the adjoint model with a final condition

$$\delta \boldsymbol{\lambda}_{K+1} = 0, \tag{4.38}$$

$$\delta \boldsymbol{\lambda}_k - \mathbf{M}_k^{i\,\mathrm{T}} \delta \boldsymbol{\lambda}_{k+1} = \mathbf{H}_k^{i\,\mathrm{T}} \mathbf{C}_{dd}^{-1}\left(\mathbf{H}^i \delta \mathbf{x} - \boldsymbol{\eta}^i\right). \tag{4.39}$$

We can now solve this coupled system efficiently using the incremental SC-4DVar (Algorithm 2).

The solution technique is similar to the one described in Sect. 4.1. We first run the nonlinear model with a first guess, $\mathbf{z}^\mathrm{f}$, to compute the model state over the whole window. This model state provides the linearization points for the tangent-linear model and observation operators and defines $\boldsymbol{\eta}^i$ and $\boldsymbol{\xi}^i$. Thus, we can evaluate the full quadratic cost function in Eq. (3.24).

After that, we compute the increment $\delta \mathbf{z}$ from the inner iterations of the linear forward model and its adjoint. We start with a first guess $\delta \mathbf{x}_0 = \delta \mathbf{z}$, which we propagate forward in time with the linear model in Eq. (4.37). This solution defines the forcing field for the backward integration of the adjoint model in Eq. (4.39). The backward integration to time $k = 0$ provides us with the gradient of the cost function in Eq. (4.35). We can then use this gradient to find a new estimate for $\delta \mathbf{z}$ by applying methods like a conjugate gradient or a quasi-Newton technique, which feeds back into the linearized forward model in Eq. (4.37).

After minimizing the linearized cost function, we add $\delta \mathbf{z}$ to the previous estimate $\mathbf{x}^i = \mathbf{x}^{i+1} + \delta \mathbf{z}$ and reevaluate $\boldsymbol{\eta}^i$ and $\boldsymbol{\xi}^i$. This update defines a new quadratic cost function and a newly updated model trajectory over the assimilation window. We then minimize this new cost function to find the next $\delta \mathbf{z}$. Thus, there are two iterations in play, one from the Gauss–Newton process, the so-called outer loop, and one set of so-called inner iterations to solve the quadratic cost function in Eq. (3.24) for each increment $\delta \mathbf{z}$. In the algorithm, γ and $\mathbf{B}$ depend on the minimization method used to solve the inner-loop problem, which can be a conjugate gradient method, BFGS, or any other minimization method.

Algorithm 2 Incremental SC-4DVar algorithm

1: Input: $\mathbf{x}_0^f \in \mathfrak{R}^n$; $\mathbf{d} \in \mathfrak{R}^m$; $\mathbf{C}_{dd} \in \mathfrak{R}^{m \times m}$; $\mathbf{C}_{zz} \in \mathfrak{R}^{n \times n}$ ▷ Prior inputs
2: $\mathbf{x}_0^0 = \mathbf{x}_0^f$ ▷ Initialization of $\mathbf{x}_0$
3: $\delta\mathbf{z} = 0$ ▷ Initialization of $\delta\mathbf{z}$
4: $i = 1$
5: **repeat**   ▷ Iteration loop
6:     $\mathbf{x}_0^i = \mathbf{x}_0^{i-1} + \delta\mathbf{z}$ ▷ Update initial condition
7:     $\boldsymbol{\xi}^i = \mathbf{x}_0^f - \mathbf{x}_0^i$ ▷ Current increment
8:     **for** $k = 0 : K$ **do**   ▷ Integrate forward model
9:        $\mathbf{x}_{k+1}^i = \mathbf{m}(\mathbf{x}_k^i)$ ▷ Nonlinear model prediction
10:     **end for**
11:     $\boldsymbol{\eta}^i = \mathbf{d} - \mathbf{h}(\mathbf{x}^i)$ ▷ Current innovation vector
12:     **repeat**   ▷ Iteration loop
13:        $\delta\mathbf{x}_0 = \delta\mathbf{z}$
14:        **for** $k = 0 : K$ **do**   ▷ Integrate forward model
15:           $\delta\mathbf{x}_{k+1} = \mathbf{M}_k^i \delta\mathbf{x}_k$
16:        **end for**
17:        $\delta\boldsymbol{\lambda}_{K+1} = 0$
18:        **for** $k = K : 0$ **do**   ▷ Integrate backward adjoint model
19:           $\delta\boldsymbol{\lambda}_k = {\mathbf{M}_k^i}^T \delta\boldsymbol{\lambda}_{k+1} - {\mathbf{H}_k^i}^T \mathbf{C}_{dd}^{-1} \left(\mathbf{H}^i \delta\mathbf{x} - \boldsymbol{\eta}^i \right)$
20:        **end for**
21:        $\delta\mathbf{z} \leftarrow \delta\mathbf{z} - \gamma\mathbf{B}\left(\delta\mathbf{z} - \boldsymbol{\xi}^i - \mathbf{C}_{zz}\delta\boldsymbol{\lambda}_0 \right)$ ▷ Update $\delta\mathbf{z}$ using gradient in Eq. (4.35)
22:     **until** convergence
23:     $i = i + 1$
24: **until** convergence

4.3   Preconditioning in Incremental SC-4DVar

As explained in Sect. 3.4, we use a Gauss–Newton method to minimize the cost function of the increments in Eq. (4.31). We can use preconditioning for fast convergence when minimizing the cost function. The most used preconditioner is a control-variable transform (Bannister, 2008, Fisher et al., 2011, Weaver et al., 2005). Remember that we define the Gauss–Newton method's ith estimate of the initial model state as $\mathbf{z}^i$ and define the new iterate as $\mathbf{z}^{i+1} = \mathbf{z}^i + \delta\mathbf{z}^i$. The preconditioning transforms the variable $\delta\mathbf{z}$ ($= \delta\mathbf{x}_0$) into $\mathbf{w}$ according to[1]

$$\delta\mathbf{z} - \boldsymbol{\xi}^i = \delta\mathbf{x}_0 - \boldsymbol{\xi}^i = \mathbf{V}\mathbf{w}^i, \qquad (4.40)$$

where $\mathbf{V} \in \mathfrak{R}^{n \times n_w}$ with $n_w \geq n$. $\mathbf{V}$ is of full rank and defined such that $\mathbf{C}_{zz} = \mathbf{V}\mathbf{V}^T$ and $\mathbf{C}_{zz}^{-1} = (\mathbf{V}^\dagger)^T\mathbf{V}^\dagger$. The superscript $\dagger$ denotes the generalized or pseudo

---

[1] An equivalent and often used transformation is to define $\delta\mathbf{z} = \mathbf{V}\mathbf{w}^i$. This alternative only slightly modifies the algorithm and leads to the same solution.

inverse. The matrix $\mathbf{V}$ typically represents the physical laws of the system, such as geostrophic balance for oceanographic simulations. We can construct $\mathbf{V}$ similarly to how we generate the background matrix in Eq. 4.7 using time-independent numerical representations of prescribed relationships between variables. For example, we might want to impose hydrostatic balance in atmospheric and ocean models, balancing the relationship between gravity and the vertical pressure gradient. The matrix $\mathbf{V}$ is then constructed such that the physical variables will follow this relation.

The transformation in Eq. (4.40) will make it possible to compute the cost function's gradients to the transformed variable $\mathbf{w} \in \mathfrak{R}^{n_w}$ without evaluating large matrices. With this formulation, the cost function Eq. (4.31) for the increments now becomes

---

Inner incremental SC-4DVar cost function with preconditioning

$$\mathcal{J}(\mathbf{w}^i) = \frac{1}{2}{\mathbf{w}^i}^{\mathrm{T}}\mathbf{w}^i + + \frac{1}{2}\left(\mathbf{H}^i \delta\mathbf{x} - \boldsymbol{\eta}^i\right)^{\mathrm{T}} \mathbf{C}_{dd}^{-1} \left(\mathbf{H}^i \delta\mathbf{x} - \boldsymbol{\eta}^i\right). \qquad (4.41)$$

It is easy to see that this control-variable transform is a form of preconditioning. If we write down the Hessian of this cost function, we find, omitting the i index for clarity,

$$\nabla_{\mathbf{w}}\nabla_{\mathbf{w}}\mathcal{J}(\mathbf{w}) = \mathbf{I} + \mathbf{V}^{\mathrm{T}}\mathbf{G}^{\mathrm{T}}\mathbf{R}^{-1}\mathbf{G}\mathbf{V}, \qquad (4.42)$$

while the Hessian of the original problem is

$$\nabla_{\delta\mathbf{z}}\nabla_{\delta\mathbf{z}}\mathcal{J}(\delta\mathbf{z}) = \mathbf{C}_{zz}^{-1} + \mathbf{G}^{\mathrm{T}}\mathbf{R}^{-1}\mathbf{G}. \qquad (4.43)$$

We immediately see that

$$\nabla_{\mathbf{w}}\nabla_{\mathbf{w}}\mathcal{J}(\mathbf{w}) = \mathbf{V}^{\mathrm{T}}\nabla_{\delta\mathbf{z}}\nabla_{\delta\mathbf{z}}\mathcal{J}(\delta\mathbf{z})\mathbf{V}. \qquad (4.44)$$

For minimizing the cost function in Eq. 4.41, it is straightforward to define a constrained minimization problem using the model definition in Eq. (4.28). We can then write the Lagrangian for this problem as

$$\begin{aligned}
\mathcal{L}(\mathbf{w}, \delta\mathbf{x}, \delta\boldsymbol{\lambda}) = & \frac{1}{2}{\mathbf{w}^i}^{\mathrm{T}}\mathbf{w}^i \\
& + \frac{1}{2}\left(\mathbf{H}^i \delta\mathbf{x} - \boldsymbol{\eta}^i\right)^{\mathrm{T}} \mathbf{C}_{dd}^{-1} \left(\mathbf{H}^i \delta\mathbf{x} - \boldsymbol{\eta}^i\right) \\
& + \sum_{k=1}^{K} \delta\boldsymbol{\lambda}_{k+1}^{\mathrm{T}}\left(\delta\mathbf{x}_{k+1}^i - \mathbf{M}_k^i \delta\mathbf{x}_k^i\right) \\
& + \left(\delta\mathbf{x}_1^i - \mathbf{M}_0^i(\boldsymbol{\xi}^i + \mathbf{V}\mathbf{w}^i)\right)^{\mathrm{T}}\delta\boldsymbol{\lambda}_1,
\end{aligned} \qquad (4.45)$$

Algorithm 3 Incremental SC-4DVar algorithm for the state-transform space

---

1: Input: $\mathbf{z}^{\mathrm{f}} \in \mathfrak{R}^n$; $\mathbf{d} \in \mathfrak{R}^m$; $\mathbf{C}_{dd} \in \mathfrak{R}^{m \times m}$; $\mathbf{V} \in \mathfrak{R}^{n \times n_w}$ ▷ Prior inputs
2:   $\mathbf{w} = 0$ ▷ Initialization
3:   $\mathbf{z}^0 = \mathbf{z}^{\mathrm{f}}$ ▷ Initialization
4:   $i = 1$ ▷ Loop initialization
5: **repeat**   ▷ Outer loop in i
6:     $\mathbf{z}^i = \mathbf{z}^{\mathrm{f}} + \mathbf{V}\mathbf{w}$ ▷ Update initial condition
7:     $\xi^i = \mathbf{z}^{\mathrm{f}} - \mathbf{z}^i$ ▷ Current increment
8:     **for** $k = 0 : K$ **do**   ▷ Integrate forward model
9:        $\mathbf{x}_{k+1}^i = \mathbf{m}(\mathbf{x}_k^i)$ ▷ Nonlinear model prediction
10:    **end for**
11:    $\eta^i = \mathbf{d} - \mathbf{h}(\mathbf{x}^i)$ ▷ Current innovation vector
12:    **repeat**   ▷ Start inner loop
13:       $\delta \mathbf{x}_0 = \xi^i + \mathbf{V}\mathbf{w}$
14:       **for** $k = 0 : K$ **do**   ▷ Integrate forward linear model
15:          $\delta \mathbf{x}_{k+1} = \mathbf{M}_k \delta \mathbf{x}_k$
16:       **end for**
17:       $\delta \lambda_{K+1} = 0$
18:       **for** $k = K : 0$ **do**   ▷ Integrate backward adjoint model
19:          $\delta \lambda_k = \mathbf{M}_k^{\mathrm{T}} \delta \lambda_{k+1} - \mathbf{H}_k^{\mathrm{T}} \mathbf{C}_{dd}^{-1} \left(\mathbf{H} \delta \mathbf{x} - \eta \right)$
20:       **end for**
21:       $\mathbf{w} \leftarrow \mathbf{w} - \gamma^i \mathbf{B}^i \mathbf{V}^{\mathrm{T}} \delta \lambda_0$ ▷ Update $\mathbf{w}$
22:    **until** convergence
23:    $i = i + 1$
24: **until** convergence

---

where the last line is just the contribution corresponding to $k = 0$ in the summation from the line above. In this transformed form, the gradient of the Lagrangian to the transformed variable $\mathbf{w}$ becomes

$$
\begin{aligned}
\nabla_{\mathbf{w}} \mathcal{L}(\mathbf{w}, \delta \mathbf{x}, \delta \lambda) &= \mathbf{w}^i - \left(\mathbf{M}_0^i \mathbf{V} \right)^{\mathrm{T}} \delta \lambda_1 \\
&= \mathbf{w}^i - \mathbf{V}^{\mathrm{T}} \mathbf{M}_0^{i\,\mathrm{T}} \delta \lambda_1 \\
&= \mathbf{w}^i - \mathbf{V}^{\mathrm{T}} \delta \lambda_0 \\
&= \mathbf{V}^{\dagger} \left(\delta \mathbf{x}_0 - \xi^i \right) - \mathbf{V}^{\mathrm{T}} \delta \lambda_0,
\end{aligned} \tag{4.46}
$$

while the gradients with respect to $\delta \mathbf{x}_k$ and $\delta \mathbf{x}_{K+1}$ are still the ones from Eqs. (4.33) and (4.34). Note that, by choosing $\mathbf{x}_0^{\mathrm{f}}$ to be the initial guess for $\mathbf{x}_0$, we do not need to transform from $\delta \mathbf{x}_0$ to $\mathbf{w}$, which would require computation of $\mathbf{V}^{\dagger}$. Thus, the gradient of the Lagrangian to $\mathbf{w}$ at time $t = 0$ becomes

$$
\nabla_{\mathbf{w}} \mathcal{L}(\mathbf{w}, \delta \lambda) = -\mathbf{V}^{\mathrm{T}} \delta \lambda_0. \tag{4.47}
$$

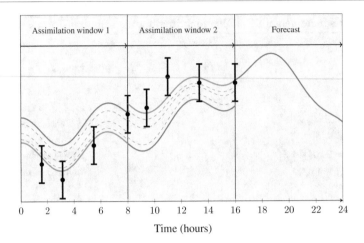

Fig. 4.1 The figure illustrates the use of incremental SC-4DVar for a case where the control vector equals the initial condition. The algorithm first integrates the model over the first assimilation window to produce the prior solution (solid blue line). After that, the SC-4DVar's inner iterations incrementally update the windows' initial condition, resulting in a sequence of updated model trajectories (dashed green lines) until convergence (solid green line), which also serves as an initial condition for the second assimilation window, and we repeat the iteration procedure

The iteration of the above equations minimizes the cost function for the linearized model system in Eq. (4.41), and we can update the state vector, i.e., the initial conditions of the nonlinear model from

$$\mathbf{z}^{i+1} = \mathbf{z}^i + \delta\mathbf{z} = \mathbf{z}^i + \boldsymbol{\xi}^i + \mathbf{V}\mathbf{w} = \mathbf{z}^{\mathrm{f}} + \mathbf{V}\mathbf{w}. \tag{4.48}$$

We can then update $\boldsymbol{\xi}^{i+1}$, run the nonlinear model from $\mathbf{z}^{i+1}$ to obtain the model solution, which we measure to compute $\boldsymbol{\eta}^{i+1}$, and start a new set of inner iterations to calculate an updated estimate of $\mathbf{w}$. Typically, we only need a few outer iterations, but we run several inner-loop iterations for each outer iteration. We provide a pseudo-code in Algorithm 3, and Fig. 4.1 illustrates the process. For more information regarding the implementation of a typical 4DVar system, see, e.g., Bannister (2017) and Weaver et al. (2005).

4.4   Summary of SC-4DVar

We have seen that the SC-4DVar method solves for the minimum of a cost function. This minimum corresponds to the maximum a posteriori (MAP) probability estimate. SC-4DVar is a gradient-based method and is thus limited to weakly nonlinear problems, as for highly nonlinear problems, the descent methods are likely to get trapped in local minima. A significant obstacle for this method is the need for a tangent linear and adjoint model, which may require a considerable if not overwhelming effort in some cases. However, so-called adjoint compilers exist and have been helpful in some cases (Marotzke et al., 1999). As one needs access to the model's code to generate the adjoint model, SC-4DVAR is not applicable with commercial "black-box" models.

The method has formed a basis for operational weather forecasting at most international weather services. The weather community has invested a massive effort in developing, maintaining, and calibrating their SC-4DVar data-assimilation systems. A particular issue with the method is that it does not provide a simple means for computing error estimates of the analysis update or propagating updated error statistics to the next assimilation window. Thus, common in the SC-4DVar systems is the use of a stationary background matrix that one designs to represent the dynamics of the model equations (Bannister, 2008).

One of the reasons for introducing the method here, except for the historical one, is that it will serve as an essential component of an ensemble SC-4DVar configuration discussed below. Finally, we have shown that it is also possible to use SC-4DVar for pure parameter estimation.

Weak Constraint 4DVar

5

It is also possible to formulate the 4DVar problem with the model acting as a weak constraint. We then search for a model solution close to the measurements that "almost" satisfies the dynamical model and its initial and boundary conditions. The concept of the model being a "weak constraint" as opposed to "strong constraint" was introduced by Sasaki (1970b). An early weak-constraint assimilation study is the one by Bennett and McIntosh (1982) who solved the weak-constraint variational inverse problem for an ocean tidal model. The two books by Bennett (1992, 2002) give a detailed presentation of the generalized weak-constraint inverse formulation and introduce a solution method known as the representer method. Below, we will discuss two approaches for including the model as a soft constraint. The first approach treats the model errors as an additional model forcing that we estimate. The second approach treats the model state over the assimilation window as the unknown variable "while allowing for model errors." It turns out that this second alternative is the easiest to solve. In the case of a nonlinear model, we follow the procedure from Sect. 3.5 where we define outer Gauss–Newton iterations and use the representer method to solve a linear inner problem for each iteration, an approach introduced by Bennett et al. (1997) and Egbert et al. (1994).

5.1 Forcing Formulation

We now assume the model system to include Eqs. (2.1, 2.4, and 2.5) and write the model as

$$\mathbf{x}_k = \mathbf{m}(\mathbf{x}_{k-1}, \mathbf{q}_k). \tag{5.1}$$

© The Author(s) 2022
G. Evensen et al., *Data Assimilation Fundamentals*,
Springer Textbooks in Earth Sciences, Geography and Environment,
https://doi.org/10.1007/978-3-030-96709-3_5

The state vector contains both initial conditions $\mathbf{x}_0$ and the time dependent model errors $\mathbf{q}_1, \ldots, \mathbf{q}_K$

$$
\mathbf{z} = \begin{pmatrix} \mathbf{x}_0 \\ \mathbf{q}_1 \\ \vdots \\ \mathbf{q}_K \end{pmatrix}, \tag{5.2}
$$

and the weak constraint cost function is again the cost function in Eq. (3.9)

The weak-constraint cost function (Forcing formulation)

$$
\mathcal{J}(\mathbf{z}) = \frac{1}{2}\left(\mathbf{z} - \mathbf{z}^{\mathrm{f}}\right)^{\mathrm{T}} \mathbf{C}_{zz}^{-1}\left(\mathbf{z} - \mathbf{z}^{\mathrm{f}}\right) + \frac{1}{2}\left(\mathbf{g}(\mathbf{z}) - \mathbf{d}\right)^{\mathrm{T}} \mathbf{C}_{dd}^{-1}\left(\mathbf{g}(\mathbf{z}) - \mathbf{d}\right), \tag{5.3}
$$

---

subject to the model constraint from Eq. (5.1).

Note that $\mathbf{C}_{zz}$ now also includes the error covariances in space and time of the model errors

$$
\mathbf{C}_{zz} = \begin{pmatrix} \mathbf{C}_{x_0 x_0} & 0 & \cdots & 0 \\ 0 & \mathbf{C}_{q_1 q_1} & \cdots & \mathbf{C}_{q_1 q_K} \\ \vdots & \vdots & \ddots & \vdots \\ 0 & \mathbf{C}_{q_K q_1} & \cdots & \mathbf{C}_{q_K q_K} \end{pmatrix}, \tag{5.4}
$$

and we naturally assume zero correlation between errors in the initial conditions and the model errors.

The operator $\mathbf{g}(\mathbf{z})$ is the composite function including the model recursion as defined in Eq. (5.1), where we apply the measurement operator to the model solution over the assimilation window. This measurement operator maps the prediction to the measurement space.

As in Sect. 4.2, we define an increment vector as

$$
\delta\mathbf{z} = \begin{pmatrix} \delta\mathbf{x}_0 \\ \delta\mathbf{q} \end{pmatrix} = \begin{pmatrix} \mathbf{x}_0^{i+1} - \mathbf{x}_0^i \\ \mathbf{q}^{i+1} - \mathbf{q}^i \end{pmatrix}. \tag{5.5}
$$

Furthermore, we use the residual $\boldsymbol{\xi}^i = \mathbf{z}^{\mathrm{f}} - \mathbf{z}^i$ from Eq. (3.23), and write the innovations $\boldsymbol{\eta}^i = \mathbf{d} - \mathbf{h}(\mathbf{z}^i)$ from Eq. (3.22), but with the state vector $\mathbf{z}$ as defined in Eq. (5.2).

In incremental WC-4DVar, we minimize the cost function in Eq. (3.24), written as

$$
\mathcal{J}(\delta\mathbf{z}) = \frac{1}{2}\left(\delta\mathbf{z} - \boldsymbol{\xi}^i\right)^{\mathrm{T}} \mathbf{C}_{zz}^{-1}\left(\delta\mathbf{z} - \boldsymbol{\xi}^i\right) + \left(\mathbf{G}^i \delta\mathbf{z} - \boldsymbol{\eta}^i\right)^{\mathrm{T}} \mathbf{C}_{dd}^{-1}\left(\mathbf{G}^i \delta\mathbf{z} - \boldsymbol{\eta}^i\right), \tag{5.6}
$$

but now $\delta\mathbf{z}$ and $\boldsymbol{\xi}^i$ also include the model error perturbations at every time step in the assimilation window. This formulation is the so-called forcing formulation (Derber, 1989; Zupanski, 1993), where we apply the model errors as a forcing of the deterministic model. It is not easy to solve the resulting Euler-Lagrange equations because of the vast dimension of $\delta\mathbf{z}$ that equals the state dimension times the number of time steps in the assimilation window. Furthermore, since any iteration method is sequential by design, there is little room for parallel computations. There have been attempts to parallelize the algorithm in the time domain. Since the cost function is quadratic in the unknowns, this is possible. However, solving this equation stands or falls by efficient preconditioning. It turns out that standard preconditioning techniques conflict with time-wise parallel computations of the problem. However, in some cases, it may be possible to represent the model errors by a lower-dimensional projection that makes the problem solvable.

5.2 State-Space Formulation

An alternative to the solving forcing formulation above is to write the weak-constraint inverse problem in Eq. (5.3) as a state-space problem. It is then required to define the model in Eq. (5.1) with additive errors

$$\mathbf{x}_k = \mathbf{m}(\mathbf{x}_{k-1}) + \mathbf{q}_k. \tag{5.7}$$

We can now replace $\mathbf{q}_k$ in Eq. (5.3) using the model definition in Eq. (5.7) to obtain

---

The weak-constraint cost function (generalized inverse formulation)

$$
\begin{aligned}
\mathcal{J}(\mathbf{x}) = {} & \frac{1}{2}\left(\mathbf{x}_0 - \mathbf{x}_0^f\right)^{\mathrm{T}} \mathbf{C}_{x_0 x_0}^{-1} \left(\mathbf{x}_0 - \mathbf{x}_0^f\right) \\
& + \frac{1}{2}\left(\mathbf{h}(\mathbf{x}) - \mathbf{d}\right)^{\mathrm{T}} \mathbf{C}_{dd}^{-1}\left(\mathbf{h}(\mathbf{x}) - \mathbf{d}\right) \\
& + \frac{1}{2}\sum_{r=1}^{K}\sum_{s=1}^{K}\left(\mathbf{x}_r - \mathbf{m}(\mathbf{x}_{r-1})\right)^{\mathrm{T}} \mathbf{C}_{qq}(r,s)\left(\mathbf{x}_s - \mathbf{m}(\mathbf{x}_{s-1})\right).
\end{aligned}
\tag{5.8}
$$

Note the double sum in the last term, which accounts for model-error correlations in time. The state vector now contains the initial conditions and the entire model solution as a discrete function of time over the data-assimilation window. In Eq. (5.8) and the following derivation, we use the notation $\mathbf{z} = \mathbf{x}$ with

$$\mathbf{x} = \begin{pmatrix} \mathbf{x}_0 \\ \vdots \\ \mathbf{x}_K \end{pmatrix}, \tag{5.9}$$

and we have eliminated the explicit appearance of the model errors $\mathbf{q}$ from the cost function.

The model-error covariance matrix is now

$$\mathbf{C}_{qq} = \begin{pmatrix} \mathbf{C}_{q_1 q_1} & \cdots & \mathbf{C}_{q_1 q_K} \\ \vdots & \ddots & \vdots \\ \mathbf{C}_{q_K q_1} & \cdots & \mathbf{C}_{q_K q_K} \end{pmatrix}, \tag{5.10}$$

and it allows for correlated errors in time, consistent with the double summation in Eq. (5.8). We note that the cost functions in Eqs. (5.3) and (5.8) are equivalent besides the assumption of additive model errors in Eq. (5.8).

5.3   Incremental Form of the Generalized Inverse

Based on the formulation in Sect. 5.2, we can formulate the incremental form of the generalized inverse, very much as in Sect. 4.2. The linearized model and measurement operators over an outer Gauss–Newton iteration increment are

$$
\begin{aligned}
\mathbf{m}\!\left(\mathbf{x}_k^{i+1}\right) &= \mathbf{m}\!\left(\mathbf{x}_k^i + \delta \mathbf{x}_k\right) \\
&\approx \mathbf{m}\!\left(\mathbf{x}_k^i\right) + \mathbf{M}_k \delta \mathbf{x}_k,
\end{aligned}
\tag{5.11}
$$

and

$$
\begin{aligned}
\mathbf{h}\!\left(\mathbf{x}^{i+1}\right) &= \mathbf{h}\!\left(\mathbf{x}^i + \delta \mathbf{x}\right) \\
&\approx \mathbf{h}\!\left(\mathbf{x}^i\right) + \mathbf{H}\delta \mathbf{x},
\end{aligned}
\tag{5.12}
$$

where $\mathbf{H}$ follows the definition from Eq. (4.12) and

$$
\mathbf{x}^{i+1} = \mathbf{x}^i + \delta \mathbf{x}.
\tag{5.13}
$$

With this, we can write the model residual in Eq. (5.8) as

$$
\begin{aligned}
\mathbf{x}_k^{i+1} - \mathbf{m}\!\left(\mathbf{x}_{k-1}^{i+1}\right) &\approx \mathbf{x}_k^{i+1} - \mathbf{m}\!\left(\mathbf{x}_{k-1}^i\right) - \mathbf{M}_{k-1}\delta \mathbf{x}_{k-1} \\
&= \mathbf{x}_k^{i+1} - \mathbf{x}_k^i + \mathbf{x}_k^i - \mathbf{m}\!\left(\mathbf{x}_{k-1}^i\right) - \mathbf{M}_{k-1}\delta \mathbf{x}_{k-1} \\
&= \delta \mathbf{x}_k - \mathbf{M}_{k-1}\delta \mathbf{x}_{k-1} + \boldsymbol{\xi}_k^i,
\end{aligned}
\tag{5.14}
$$

where for the time steps $k = 1, \dots, K$

$$
\boldsymbol{\xi}_k^i = \mathbf{x}_k^i - \mathbf{m}\!\left(\mathbf{x}_{k-1}^i\right),
\tag{5.15}
$$

is the deviation of the model trajectory from the exact model solution. For the increment's initial-condition term we define as before $\boldsymbol{\xi}_0^i = \mathbf{x}_0^i - \mathbf{x}_0^f$. The innovations are

$$
\boldsymbol{\eta}^i = \mathbf{d} - \mathbf{h}\!\left(\mathbf{x}^i\right).
\tag{5.16}
$$

We insert these definitions into the cost function for the generalized inverse in Eq. (5.8) to get the inner cost function as

---

The weak-constraint incremental cost function

$$
\begin{aligned}
\mathcal{J}(\delta \mathbf{x}) ={}& \frac{1}{2}\left(\delta \mathbf{x}_0 - \boldsymbol{\xi}_0^i\right)^{\mathrm{T}} \mathbf{C}_{x_0 x_0}^{-1}\left(\delta \mathbf{x}_0 - \boldsymbol{\xi}_0^i\right) \\
&+ \frac{1}{2}\left(\mathbf{H}^i \delta \mathbf{x} - \boldsymbol{\eta}^i\right)^{\mathrm{T}} \mathbf{C}_{dd}^{-1}\left(\mathbf{H}^i \delta \mathbf{x} - \boldsymbol{\eta}^i\right) \\
&+ \frac{1}{2}\sum_{r=1}^{K}\sum_{s=1}^{K}\left(\delta \mathbf{x}_r - \mathbf{M}_{r-1}\delta \mathbf{x}_{r-1} + \boldsymbol{\xi}_r^i\right)^{\mathrm{T}} \mathbf{C}_{qq}^{-1}(r, s)\left(\delta \mathbf{x}_s - \mathbf{M}_{s-1}\delta \mathbf{x}_{s-1} + \boldsymbol{\xi}_s^i\right).
\end{aligned}
\tag{5.17}
$$

Since the model is linear in the inner loop, a Gaussian prior for the initial condition leads to a Gaussian prior on the model states in the whole time window. Hence, we can again assume a Gaussian prior for the unknown state increment $\delta \mathbf{x} = \mathbf{x}^{i+1} - \mathbf{x}^i$ for the model's initial condition and the model solution at all instants in the assimilation window.

5.4 Minimizing the Cost Function for the Increment

Taking the gradient of the cost function in Eq. (5.17) to the model state $\delta \mathbf{x}_k$ gives, for $k \neq 0$

$$
\begin{aligned}
\nabla_{\delta \mathbf{x}_k} \mathcal{J}(\delta \mathbf{x}) = & \, \mathbf{H}_k^{i\,\mathrm{T}} \mathbf{C}_{dd}^{-1} \left(\mathbf{H}^i \delta \mathbf{x} - \boldsymbol{\eta}^i \right) \\
& + \sum_{s=1}^{K} \mathbf{C}_{qq}^{-1}(k, s) \left(\delta \mathbf{x}_s - \mathbf{M}_{s-1}^i \delta \mathbf{x}_{s-1} + \boldsymbol{\xi}_s^i \right) \\
& - \mathbf{M}_k^{i\,\mathrm{T}} \sum_{s=1}^{K} \mathbf{C}_{qq}^{-1}(k+1, s) \left(\delta \mathbf{x}_s - \mathbf{M}_{s-1}^i \delta \mathbf{x}_{s-1} + \boldsymbol{\xi}_s^i \right).
\end{aligned}
\tag{5.18}
$$

From this equation, we can define an adjoint vector for each time step as

$$
\delta \boldsymbol{\lambda}_k = \sum_{s=1}^{K} \mathbf{C}_{qq}^{-1}(k, s) \left(\delta \mathbf{x}_s - \mathbf{M}_{s-1}^i \delta \mathbf{x}_{s-1} + \boldsymbol{\xi}_s^i \right),
\tag{5.19}
$$

such that we can write Eq. (5.18) as

$$
\nabla_{\delta \mathbf{x}_k} \mathcal{J}(\delta \mathbf{x}) = \mathbf{H}_k^{i\,\mathrm{T}} \mathbf{C}_{dd}^{-1} \left(\mathbf{H}^i \delta \mathbf{x} - \boldsymbol{\eta}^i \right) + \delta \boldsymbol{\lambda}_k - \mathbf{M}_k^{i\,\mathrm{T}} \delta \boldsymbol{\lambda}_{k+1}.
\tag{5.20}
$$

For the initial time $k = 0$ we find for the gradient

$$
\begin{aligned}
\nabla_{\delta \mathbf{x}_0} \mathcal{J}(\delta \mathbf{x}) &= \mathbf{C}_{x_0 x_0}^{-1} \left(\delta \mathbf{x}_0 - \boldsymbol{\xi}_0^i \right) - \mathbf{M}_0^{i\,\mathrm{T}} \delta \boldsymbol{\lambda}_1 \\
&= \mathbf{C}_{x_0 x_0}^{-1} \left(\delta \mathbf{x}_0 - \boldsymbol{\xi}_0^i \right) - \delta \boldsymbol{\lambda}_0,
\end{aligned}
\tag{5.21}
$$

where we used the definition of the adjoint variable (5.19).

If we now set all gradients of the cost function to zero, we find the Euler-Lagrange equations, which comprise a two-point boundary value problem in time consisting of the forward model with the initial condition for the increments $\delta \mathbf{x}$

Forward model

$$
\delta \mathbf{x}_0 = \boldsymbol{\xi}_0^i + \mathbf{C}_{x_0 x_0} \delta \boldsymbol{\lambda}_0,
\tag{5.22}
$$

$$
\delta \mathbf{x}_k - \mathbf{M}_{k-1}^i \delta \mathbf{x}_{k-1} = -\boldsymbol{\xi}_k^i + \sum_{s=1}^{K} \mathbf{C}_{qq}(k, s) \delta \boldsymbol{\lambda}_s,
\tag{5.23}
$$

---

and the backward model for the adjoint variable $\delta \boldsymbol{\lambda}$

Backward model

$$\delta\lambda_{K+1} = \mathbf{0}, \tag{5.24}$$

$$\delta\lambda_k - \mathbf{M}_k^{i\,\mathrm{T}}\delta\lambda_{k+1} = -\mathbf{H}_k^{i\,\mathrm{T}}\mathbf{C}_{dd}^{-1}\left(\mathbf{H}^i\delta\mathbf{x} - \boldsymbol{\eta}^i\right). \tag{5.25}$$

The Eqs. (5.22)–(5.25) define the minimizing solution of the variational problem defined by the incremental cost function in Eq. (5.17). Due to the coupling of these equations, an iterative solution procedure is a natural choice. Note that by setting $\mathbf{C}_{qq} = 0$, we decouple the forward model integration from the adjoint variable, leading to the SC-4DVar method discussed above. In this case, we restrict ourselves to iteratively solving for the initial conditions as the only unknown. As an alternative to iterative solution methods, the representer method decouples the forward and backward models. We will discuss this approach in the following section.

5.5   Observation Space Formulation

In the following, we explore that the observation space is typically much smaller than the state space, which is even more true for the weak-constraint case due to the larger state vector. While the problem size grows dramatically in the state-space from the strong-constraint formulation to the weak-constraint 4DVar, it does not grow in the observation space. Hence, solving the weak-constraint problem in observation space is likely more efficient, as was realized and discussed by Bennett (1992). He formulated a solution method for the weak-constraint problem called the representer method. While the original representer method is illustrative, it is not efficient with many observations, as the method requires a backward (adjoint) and forward model integration for each measurement. A later variant by Egbert et al. (1994) avoids this problem. We will discuss the representer method below as it is highly efficient for linear inverse problems and provides for further insight into the data assimilation problem.

5.5.1   Original Representer Method

The representer method by Bennett (1992) exploits the "weak coupling" in the Euler-Lagrange Eqs. (5.22)–(5.25) through the measurement term in Eq. (5.24). Let's assume a solution of the form

$$\delta\mathbf{x} = \delta\mathbf{x}^{\mathrm{f}} + \sum_{p=1}^{m} b_p \mathbf{r}_p = \delta\mathbf{x}^{\mathrm{f}} + \mathbf{R}\mathbf{b}, \tag{5.26}$$

$$\delta\boldsymbol{\lambda} = \delta\boldsymbol{\lambda}^{\mathrm{f}} + \sum_{p=1}^{m} b_p \mathbf{s}_p = \delta\boldsymbol{\lambda}^{\mathrm{f}} + \mathbf{S}\mathbf{b}. \tag{5.27}$$

Here we assume $\delta \mathbf{x}^{\mathrm{f}} \neq 0$ and $\delta \boldsymbol{\lambda}^{\mathrm{f}} = 0$ to be a first-guess solution which would result from the case with no "observations", i.e., no forcing term in Eq. (5.24). In Eqs. (5.26) and (5.27) we assume the solution to equal the first guess plus a linear combination of m representers or influence functions $\mathbf{r}_p$ and their adjoints $\mathbf{s}_p$. There is one representer function for each of the m measurements, and we store them in the m columns of the matrix $\mathbf{R}$. Bennett (1992) showed that this linear combination of representer functions exactly represents the minimizing solution.

Inserting the expressions for $\delta \mathbf{x}$ and $\delta \boldsymbol{\lambda}$ in the Euler-Lagrange Eqs. (5.22)–(5.25) gives for the first-guess solution

$$\delta \mathbf{x}_0^{\mathrm{f}} = \boldsymbol{\xi}_0^i, \tag{5.28}$$

$$\delta \mathbf{x}_k^{\mathrm{f}} - \mathbf{M}_{k-1}^i \delta \mathbf{x}_{k-1}^{\mathrm{f}} = -\boldsymbol{\xi}_k^i, \tag{5.29}$$

$$\delta \boldsymbol{\lambda}_{K+1}^{\mathrm{f}} = 0, \tag{5.30}$$

$$\delta \boldsymbol{\lambda}_k^{\mathrm{f}} - \mathbf{M}_k^{i\,\mathrm{T}} \delta \boldsymbol{\lambda}_{k+1}^{\mathrm{f}} = 0. \tag{5.31}$$

For the representers and their adjoints we obtain

$$\mathbf{R}_0 \mathbf{b} = \mathbf{C}_{x_0 x_0} \mathbf{M}_0^{i\,\mathrm{T}} \mathbf{S}_1 \mathbf{b}, \tag{5.32}$$

$$\left(\mathbf{R}_k \mathbf{b} - \mathbf{M}_{k-1}^i \mathbf{R}_{k-1} \mathbf{b} \right) = \sum_{s=1}^{K} \mathbf{C}_{qq}(k, s) \mathbf{S}_s \mathbf{b}, \tag{5.33}$$

$$\mathbf{S}_{K+1} \mathbf{b} = \mathbf{0}, \tag{5.34}$$

$$\mathbf{S}_k \mathbf{b} - \mathbf{M}_k^{i\,\mathrm{T}} \mathbf{S}_{k+1} \mathbf{b} = \mathbf{H}_k^{i\,\mathrm{T}} \mathbf{C}_{dd}^{-1} \left(\boldsymbol{\eta}^i - \mathbf{H}^i (\delta \mathbf{x}^{\mathrm{f}} + \mathbf{R} \mathbf{b}) \right). \tag{5.35}$$

The decomposition in Eqs.(5.26) and (5.27) enforce that $\mathbf{R}$ and $\mathbf{S}$ are not functions of $\mathbf{b}$. This can be achieved by defining $\mathbf{b}$ as

$$\mathbf{b} = \mathbf{C}_{dd}^{-1} \left(\boldsymbol{\eta}^i - \mathbf{H}^i (\delta \mathbf{x}^{\mathrm{f}} + \mathbf{R} \mathbf{b}) \right), \tag{5.36}$$

such that Eq. (5.35) simplifies to

$$\mathbf{S}_k \mathbf{b} - \mathbf{M}_k^{i\,\mathrm{T}} \mathbf{S}_{k+1} \mathbf{b} = \mathbf{H}_k^{i\,\mathrm{T}} \mathbf{b}. \tag{5.37}$$

Since $\mathbf{b} \neq 0$ and acts as a common multiplier in all the Eqs. (5.32, 5.33, 5.34, and 5.37) we can write the following uncoupled system of equations for the representers and their adjoints,

$$\mathbf{R}_0 = \mathbf{C}_{x_0 x_0} \mathbf{M}_0^{i\,\mathrm{T}} \mathbf{S}_1, \tag{5.38}$$

$$\mathbf{R}_k - \mathbf{M}_{k-1}^i \mathbf{R}_{k-1} = \sum_{s=1}^{K} \mathbf{C}_{qq}(k, s) \mathbf{S}_s, \tag{5.39}$$

$$\mathbf{S}_{K+1} = \mathbf{0}, \tag{5.40}$$

$$\mathbf{S}_k - \mathbf{M}_k^{i\,\mathrm{T}} \mathbf{S}_{k+1} = \mathbf{H}_k^{i\,\mathrm{T}}. \tag{5.41}$$

Here $\mathbf{H}_k^i$ is the columns of $\mathbf{H}^i$ in Eq. (4.12) corresponding to the time k. The matrix $\mathbf{R}$'s columns contain the influence functions $\mathbf{r}_s$, and the matrix $\mathbf{S}$ contains their adjoints.

Thus, a backward-in-time integration of Eqs. (5.40) and (5.41) determines $\mathbf{S}$ and we can then solve for $\mathbf{R}$ by a forward integration of Eqs. (5.38) and (5.39). What remains is then to determine $\mathbf{b}$ from Eq. (5.36), i.e.,

---

Linear system for the representer coefficients b

$$\left(\mathbf{H}^i\mathbf{R} + \mathbf{C}_{dd}\right)\mathbf{b} = \eta^i - \mathbf{H}^i\delta\mathbf{x}^{\mathrm{f}}. \tag{5.42}$$

Bennett (2002) gives a detailed explanation of how to solve for the representer solution efficiently. First, note that as soon as we have computed $\mathbf{b}$ by solving Eq. (5.42), we can use the definition of $\mathbf{b}$ from Eq. (5.36) and write the adjoint Eq. (5.25) as

---

Adjoint equation forced by b

$$\delta\lambda_k - \mathbf{M}_k^{i\,\mathrm{T}}\delta\lambda_{k+1} = \mathbf{H}_k^{i\,\mathrm{T}}\mathbf{b}. \tag{5.43}$$

Thus, knowing $\mathbf{b}$ allows us to compute the solution by one backward integration of Eq. (5.43) subject to the final condition in Eq. (5.24), followed by one forward integration of the model with initial condition in Eqs. (5.22) and (5.23).

Notice that we do not need to store all the representers and their adjoints. We must only construct the "representer matrix"

$$\mathcal{R} = \mathbf{HR} \tag{5.44}$$

that enables us to solve the system in Eq. (5.42).

The representer method has a beautiful property. It shows via its basic construction in Eq. (5.26) that the solution to the linear data-assimilation problem is the first-guess solution plus a linear combination of the representers. Furthermore, we can interpret any representer $\mathbf{r}_p$ as the influence function for measurement p. We construct it as column p of the matrix Eqs. (5.38)–(5.41). Specifically, each measurement generates a forcing field for the adjoint representer at the observation time. This forcing field is propagated back to the initial time via the adjoint representer equations. After that, the adjoint representers $\mathbf{s}_p$ force the forward integration when computing the representers $\mathbf{r}_p$. In this way, we spread out the influence of measurement p over the whole space-time domain. And, as mentioned, the complete solution to the linear problem is the first guess plus a linear combination of these space-time influence functions, one for each measurement.

The solution method outlined above is inefficient as it needs a full adjoint and forward model integration for each observation. On the other hand, the solution method illustrates well the nature of the weak-constraint estimation problem. Note also that the representer solution is the unique minimizing solution of the Euler-Lagrange equations, as first noticed by Bennett (1992). The following section discusses a much more efficient implementation of the representer method. For an application that illustrates some of the representer method's properties, we refer to the example in Chap. 17.

5.5.2 Efficient Weak-Constraint Solution in Observation Space

In the previous section, we saw that we need to solve for the vector $\mathbf{b}$ from Eq. (5.42), which we write as

$$\left(\mathcal{R} + \mathbf{C}_{dd}\right)\mathbf{b} = \eta^i - \mathbf{H}^i\delta\mathbf{x}^f, \tag{5.45}$$

using the definition of the representer matrix in Eq. (5.44).

Recall that, as soon as $\mathbf{b}$ is known, we can find the final solution via one backward integration of Eq. (5.43) subject to the condition in Eq. (5.24), followed by an integration of the model in Eqs. (5.22) and (5.23).

Apparently, the definition in Eq. (5.44) requires us to form the representer matrix $\mathcal{R}$ to solve the linear system in Eq. (5.45), for which we need to compute all the representers. However, suppose we use an iterative method for solving Eq. (5.45). In that case we only need to calculate products $(\mathcal{R} + \mathbf{C}_{dd})\mathbf{v}$ for arbitrary vectors $\mathbf{v}$. As $\mathbf{C}_{dd}$ is known, the problem is reduced to computing the product $\mathcal{R}\mathbf{v} = \mathbf{HR}\mathbf{v}$.

But, if we write the Eqs. (5.32), (5.33), (5.34), and (5.37) as

$$(\mathbf{R}_0\mathbf{v}) = \mathbf{C}_{x_0x_0}\mathbf{M}_0^{i\,\mathrm{T}}(\mathbf{S}_1\mathbf{v}), \tag{5.46}$$

$$(\mathbf{R}_k\mathbf{v}) = \mathbf{M}_{k-1}^i(\mathbf{R}_{k-1}\mathbf{v}) + \sum_{s=1}^{K}\mathbf{C}_{qq}(k,s)(\mathbf{S}_s\mathbf{v}), \tag{5.47}$$

$$(\mathbf{S}_{K+1}\mathbf{v}) = \mathbf{0}, \tag{5.48}$$

$$(\mathbf{S}_k\mathbf{v}) = \mathbf{M}_k^{i\,\mathrm{T}}(\mathbf{S}_{K+1}\mathbf{v}) + \mathbf{H}_k^{i\,\mathrm{T}}\mathbf{v}, \tag{5.49}$$

we can compute the product $\mathbf{c} = \mathbf{R}\mathbf{v}$ by one backward integration of Eq. (5.49) subject to the final condition (5.48) to obtain the field $\boldsymbol{\psi} = \mathbf{S}\mathbf{v}$, followed by a forward integration of the Eq. (5.47) from the initial condition in Eq. (5.46). Thus, we rewrite these equations as

$$\mathbf{c}_0 = \mathbf{C}_{x_0x_0}\mathbf{M}_0^{i\,\mathrm{T}}\boldsymbol{\psi}_1, \tag{5.50}$$

$$\mathbf{c}_k = \mathbf{M}_{k-1}^i\mathbf{c}_{k-1} + \sum_{s=1}^{K}\mathbf{C}_{qq}(k,s)\boldsymbol{\psi}_s, \tag{5.51}$$

$$\boldsymbol{\psi}_{K+1} = \mathbf{0}, \tag{5.52}$$

$$\boldsymbol{\psi}_k = \mathbf{M}_k^{i\,\mathrm{T}}\boldsymbol{\psi}_{k+1} + \mathbf{H}_k^{i\,\mathrm{T}}\mathbf{v}. \tag{5.53}$$

By measuring this solution, we find for any nonzero vector $\mathbf{v}$

$$\mathbf{Hc} = \mathbf{HR}\mathbf{v} = \mathcal{R}\mathbf{v}, \tag{5.54}$$

which is precisely the matrix-vector product we need to compute.

This algorithm by Egbert et al. (1994) calculates the product $\mathcal{R}\mathbf{v}$ without knowing $\mathcal{R}$ by performing one backward and one forward model integration. Thus, it is possible to solve Eq. (5.45) iteratively for $\mathbf{b}$, using only two model integrations per iteration. *Isn't that an astonishing result?* In the following two sections, we introduce an iterative solver to illustrate the methodology. We also discuss an efficient approach for computing the convolutions of the model- and initial-error covariances with the adjoint variable.

We present an example algorithm for solving the weak-constraint inverse problem in Algorithm 4. Here we combine the outer incremental Gauss–Newton iterations with the representer method for solving the linear inverse problems for the increments. Note that it is also possible to use the Algorithm 4 to solve the strong constraint variational problem by setting $\mathbf{C}_{qq} \equiv 0$.

5.5.2.1   Iterative Equation Solver

We can illustrate the iterative procedure for solving Eq. (5.45) by using a steepest descent method. Let's write the linear system in Eq. (5.45) as

$$\mathbf{C}\mathbf{b} = \boldsymbol{\eta}. \tag{5.55}$$

Solving Eq. (5.55) is equivalent to minimizing the functional

$$\phi(\mathbf{b}) = \frac{1}{2}\mathbf{b}^{\mathrm{T}}\mathbf{C}\mathbf{b} - \mathbf{b}^{\mathrm{T}}\boldsymbol{\eta}, \tag{5.56}$$

which has a gradient

$$\nabla_{\mathbf{b}}\phi(\mathbf{b}) = \boldsymbol{\rho} = \mathbf{C}\mathbf{b} - \boldsymbol{\eta}. \tag{5.57}$$

We can then minimize the cost function in Eq. (5.56) iteratively by using an iterative approach, .e.g.,

$$\mathbf{b}^{i+1} = \mathbf{b}^{i} - \gamma\boldsymbol{\rho}, \tag{5.58}$$

where $\gamma = \boldsymbol{\rho}^{\mathrm{T}}\boldsymbol{\rho}/\boldsymbol{\rho}^{\mathrm{T}}\mathbf{C}\boldsymbol{\rho}$ is the optimal steepest-descent steplength. Of course, in real problems we should introduce preconditioning or use a conjugate gradient method to speed up the convergence.

5.5.2.2   Fast Computation of the Error Terms

The final issue of using the representer method for real-sized problems, is how to compute the discrete convolutions in the model error term in Eqs. (5.23) and (5.51), and in Algorithm 4. The direct computation of matrix-vector multiplications in these terms becomes too computationally demanding for real problems. Even for the initial conditions, the multiplication $\mathbf{C}_{x_0 x_0}\delta\boldsymbol{\lambda}_0$ requires n^2 operations, with n being the size of the model state vector.

The approach taken by Bennett (1992) was to use a Gaussian covariance where the Fourier transform is known and diagonal and use fast Fourier transforms to compute the multiplication efficiently in Fourier space at a cost proportional to $n \ln n$. Bennett (2002) pointed out that with certain assumptions on the "shape" of the covariance matrix $\mathbf{C}_{qq}(k, s)$, even more efficient methods exist.

Let's assume the covariance to be isotropic and separable in space and time with Gaussian covariance in space and exponential covariance in time, i.e.,

$$\mathbf{C}_{qq}(x_i, x_j, t_r, t_s) \propto \exp\left(-|x_i - x_j|^2/r_x^2\right) \exp\left(-|t_r - t_s|/\tau\right), \tag{5.59}$$

Algorithm 4 Incremental WC-4DVar using the representer method

---

1: Input: $\mathbf{x}_0^{\mathrm{f}} \in \mathfrak{R}^n$; $\mathbf{d} \in \mathfrak{R}^m$; $\mathbf{C}_{dd} \in \mathfrak{R}^{m \times m}$; $\mathbf{C}_{x_0 x_0} \in \mathfrak{R}^{n \times n}$; $\mathbf{C}_{qq}(r, s) \in \mathfrak{R}^{n \times n}$ ▷ Prior inputs

    ▷ Forward integration of nonlinear model to get prior solution $\mathbf{x}$ for the window

2: **for** $k = 0 : K$ **do**

3:     $\mathbf{x}_{k+1}^{\mathrm{f}} = \mathbf{m}(\mathbf{x}_k^{\mathrm{f}})$ ▷ Eq. (5.7) with $\mathbf{q}_k = 0$

4: **end for**

    ▷ Outer incremental GN iterations

5: $i = 1$; $\mathbf{x}^i = \mathbf{x}^{\mathrm{f}}$; $\mathbf{b} = 0$

6: **repeat**

7:     $\boldsymbol{\xi}_0^i = \mathbf{x}_0^{\mathrm{f}} - \mathbf{x}_0^i$

8:     **for** $k = 1 : K$ **do**

9:        $\boldsymbol{\xi}_k^i = \mathbf{x}_k^i - \mathbf{m}(\mathbf{x}_{k-1}^i)$ ▷ Eq. (5.15)

10:     **end for**

11:     $\boldsymbol{\eta}^i = \mathbf{d} - \mathbf{h}(\mathbf{x}^i)$ ▷ Eq. (5.16)

    ▷ Inner iterative representer solution solving for $\mathbf{b}$: $\left(\mathcal{R} + \mathbf{C}_{dd}\right)\mathbf{b} = \boldsymbol{\eta}^i - \mathbf{H}^i \delta\mathbf{x}$

12:     **repeat**

13:        $\psi_{K+1} = 0$ ▷ Eq. (5.52)

14:        **for** $k = K : 0$ **do**

15:           $\psi_k = \mathbf{M}_k^{i^{\mathsf{T}}} \psi_{k+1} + \mathbf{H}_k^{i^{\mathsf{T}}} \mathbf{b}$ ▷ Eq. (5.53)

16:        **end for**

17:        $\mathbf{c}_0^i = \mathbf{C}_{x_0 x_0} \psi_0$ ▷ Eq. (5.50)

18:        **for** $k = 1 : K$ **do**

19:           $\mathbf{c}_k = \mathbf{m}(\mathbf{c}_{k-1}) + \sum_{s=1}^{K} \mathbf{C}_{qq}(k, s)\psi_s$ ▷ Eq. (5.51)

20:        **end for**

21:        $\boldsymbol{\rho} = \mathbf{H}^i\mathbf{c} + \mathbf{C}_{dd}\mathbf{b} - \mathbf{H}^i \delta\mathbf{x}^{\mathrm{fg}} + \boldsymbol{\eta}^i$ ▷ Eq. (5.57)

22:        $\mathbf{b} = \mathbf{b} - \gamma\boldsymbol{\rho}$ ▷ Eq. (5.58)

23:     **until** $\boldsymbol{\rho} \approx 0$

    ▷ Backward integration to obtain solution given estimate of $\mathbf{b}$

24:     $\delta\boldsymbol{\lambda}_{K+1} = 0$ ▷ Eq. (5.24)

25:     **for** $k = K : 0$ **do**

26:        $\delta\boldsymbol{\lambda}_k = \mathbf{M}_k^{i^{\mathsf{T}}} \delta\boldsymbol{\lambda}_{k+1} + \mathbf{H}_k^{i^{\mathsf{T}}} \mathbf{b}$ ▷ Eq. (5.43)

27:     **end for**

    ▷ Forward integration to obtain solution given estimate of $\mathbf{b}$

28:     $\delta\mathbf{x}_0 = \boldsymbol{\xi}_0^i + \mathbf{C}_{x_0 x_0} \delta\boldsymbol{\lambda}_0$ ▷ Eq. (5.22)

29:     **for** $k = 0 : K$ **do**

30:        $\delta\mathbf{x}_{k+1} = \mathbf{M}_k^i \delta\mathbf{x}_k - \boldsymbol{\xi}_k^i + \sum_{s=1}^{K} \mathbf{C}_{qq}(k, s)\delta\boldsymbol{\lambda}_s$ ▷ Eq. (5.23)

31:     **end for**

32:     $i = i + 1$

33:     $\mathbf{x}^i = \mathbf{x}^{i-1} + \delta\mathbf{x}$ ▷ Eq. (5.13)

34: **until** $\delta\mathbf{x} \approx \mathbf{0}$

---

where x_i and x_j are two spatial model gridpoints, and t_r and t_s are two time steps. We also define the decorrelation lengths r_x in space and τ in time. Then, for a model with spatial dimension d we must compute the following

$$\delta\lambda_{xt}(i, r) = \sum_{j=1}^{n}\sum_{s=1}^{K} \exp\left(-|x_i - x_j|^2/r_x^2\right) \exp\left(-|t_r - t_s|/\tau\right) \delta\lambda(j, s) \qquad (5.60)$$

$$= \sum_{j=1}^{n} \exp\left(-|x_i - x_j|^2/r_x^2\right) \sum_{s=1}^{K} \exp\left(-|t_r - t_s|/\tau\right) \delta\lambda(j, s) \qquad (5.61)$$

$$= \sum_{j=1}^{n} \exp\left(-|x_i - x_j|^2/r_x^2\right) \delta\lambda_t(j, r), \qquad (5.62)$$

where we defined

$$\delta\lambda_t(j, r) = \sum_{s=1}^{K} \exp\left(-|t_r - t_s|/\tau\right) \delta\lambda(j, s). \qquad (5.63)$$

The solution procedure proposed by Bennett (2002) uses that the expression in Eq. (5.63) when written in the continuous form in time

$$\delta\lambda_t(x_j, t) = \int_0^T \exp\left(-|t - t'|/\tau\right) \delta\lambda\left(x_j, t'\right) dt', \qquad (5.64)$$

for each value of x_j, is the solution of a two-point boundary value problem in time (Bennett 2002, see pages 65–66),

$$\frac{\partial^2 \delta\lambda_t}{\partial t^2} - \frac{1}{\tau^2}\delta\lambda_t = -\frac{2\delta\lambda(x_j, \tau)}{\tau} \qquad (5.65)$$

$$\frac{\partial \delta\lambda_t}{\partial t} - \frac{1}{\tau}\delta\lambda_t = 0 \quad \text{for } t = 0 \qquad (5.66)$$

$$\frac{\partial \delta\lambda_t}{\partial t} + \frac{1}{\tau}\delta\lambda_t = 0 \quad \text{for } t = T. \qquad (5.67)$$

We can solve this one-dimensional boundary value problem for each of the $j = 1, n$ functions $\delta\lambda(x_j, \tau)$ to obtain $\delta\lambda_t(j, r)$.

Furthermore, as soon as we have $\delta\lambda_t(x, t)$, or in discrete form $\delta\lambda_t(j, r)$, then the expression in Eq. (5.62), becomes in continous form

$$\delta\lambda_{xt}(x, t) = \int_{-\infty}^{\infty} \exp\left(-|x - x'|^2/r_x^2\right) \delta\lambda_t\left(x', t\right) dx'. \qquad (5.68)$$

We will now use that the variable $\theta(x, s)$ defined as

$$\theta(x, s) = (4\pi s)^{-\frac{d}{2}} \int_{-\infty}^{\infty} \exp\left(-|x - x'|^2/(4s)\right) \delta\lambda_t\left(x', s\right) dx'. \qquad (5.69)$$

is the solution of the heat equation

$$\frac{\partial \theta}{\partial s} = \nabla^2 \theta, \qquad (5.70)$$

subject to the initial condition

$$\theta(x, s = 0) = \delta\lambda_t(x, t),\tag{5.71}$$

(Bennett, 2002; Wikipedia, 2022). Thus, $\delta\lambda_{xt}(x, t)$ becomes

$$\delta\lambda_{xt}(x, t) = \left(\pi r_x^2\right)^{\frac{d}{2}} \theta\left(x, s = r_x^2/4\right),\tag{5.72}$$

so, we just need to integrate Eq. (5.70) from $s = 0$ to $s = r_x^2/4$ to get $\theta(s)$ and hence $\delta\lambda_{xt}(x, t)$ from Eq. (5.72).

For the initial conditions, we can compute the convolution in Eq. (5.68) at a cost proportional to n, or precisely n times the required number of pseudo time-steps needed when solving the diffusion Eq. (5.70) from $s = 0$ to $s = r_x^2/4$. We need to solve the diffusion Eq. (5.70) for each time step for the model error term. Thus the computational cost becomes proportional to nK. Additionally, we need to solve the boundary value problem in Eqs. (5.65)–(5.67) nK times, one time for each model gridpoint in space and time.

Bennett (2002) provides a detailed description of this algorithm, and we note that Courtier (1997) also discusses the incremental weak constraint method in conjunction with WC-4DVar. The US Navy uses an operational implementation of the representer method with an ocean model (Souopgui et al., 2017). They also have a dormant representer implementation for their atmospheric model.

Kalman Filters and 3DVar

6

This chapter discusss solution methods for particular cases of the minimization problem defined by the cost function in Eq. (3.9). We start by looking for a closed-form solution that minimizes the cost function, and then we continue discussing how specific cases lead to several well-known methods. The first case assumes that the measurements are all located at the initial time of the assimilation window. Thus, there is no need for any model integrations during the minimization. The problem then reduces to the classical *3-dimensional variational (3DVar)* formulation. The second case assumes that the model and the measurement operator are linear, allowing us to find an explicit gradient solution in Eq. (3.11). This particular case leads to the *Kalman filter (KF)* update equations. And, additionally, if we have measurements located at the initial time of the assimilation window, we obtain the standard form of the KF. To further simplify the KF, we get the simplified optimal interpolation (OI) algorithm by ignoring the time evolution of error statistics. In addition to these specific methods, we consider the weakly nonlinear case where we can sometimes still use the Kalman filter equations with its linearized model and measurement operator in the *extended Kalman filter (EKF)*.

6.1 Linear Update from Predicted Measurements

To explore possible linear solutions to the estimation problem, let's start from a closed-form solution of the estimation problem in Eq. (3.11) in the trivial case when $\mathbf{g}(\mathbf{z}) = \mathbf{G}\mathbf{z}$ is linear. We assume that the state vector is the model solution at the initial time of the assimilation window $\mathbf{z} = \mathbf{x}_0$, and we have the measurements distributed over the assimilation window. In this case, Eq. (3.11) becomes

$$\mathbf{C}_{zz}^{-1}\left(\mathbf{z}^{a} - \mathbf{z}^{f}\right) + \mathbf{G}^{T}\mathbf{C}_{dd}^{-1}\left(\mathbf{G}\mathbf{z}^{a} - \mathbf{d}\right) = 0, \tag{6.1}$$

© The Author(s) 2022
G. Evensen et al., *Data Assimilation Fundamentals*,
Springer Textbooks in Earth Sciences, Geography and Environment,
https://doi.org/10.1007/978-3-030-96709-3_6

which has an explicit solution

$$\mathbf{z}^{a} = \mathbf{z}^{f} + \left(\mathbf{C}_{zz}^{-1} + \mathbf{G}^{T}\mathbf{C}_{dd}^{-1}\mathbf{G}\right)^{-1}\mathbf{G}^{T}\mathbf{C}_{dd}^{-1}\left(\mathbf{d} - \mathbf{G}\mathbf{z}^{f}\right). \tag{6.2}$$

Here $\mathbf{y} = \mathbf{G}\mathbf{z}$ is the linear prediction of the measurements, which we can write as

$$\mathbf{y} = \mathbf{G}\mathbf{z} = \mathbf{H}\begin{pmatrix} \mathbf{z} \\ \mathbf{x}_1 \\ \vdots \\ \mathbf{x}_K \end{pmatrix} = \mathbf{H}\begin{pmatrix} \mathbf{z} \\ \mathbf{M}_1\mathbf{z} \\ \vdots \\ \mathbf{M}_K\ldots\mathbf{M}_1\mathbf{z} \end{pmatrix} = \mathbf{H}\begin{pmatrix} \mathbf{I} \\ \mathbf{M}_1 \\ \vdots \\ \mathbf{M}_K\ldots\mathbf{M}_1 \end{pmatrix}\mathbf{z} = \mathbf{H}\mathcal{M}\mathbf{z}, \tag{6.3}$$

which defines $\mathbf{G}$, and where we introduce $\mathcal{M}$ for later use. Thus, if the model is linear, we can compute the update at the initial time of the assimilation window from measurements located throughout the assimilation window by solving Eq. (6.2). We include a time-step index on $\mathbf{M}$ so that the equation will also apply in the nonlinear case where $\mathbf{M}_k$ is the tangent-linear model at time t_k.

To find an explicit solution of the estimation problem in Eq. (3.11) in the nonlinear case, we introduce the following approximation

---

Approximation 5 (Linearization)   *Linearize $\mathbf{g}(\mathbf{z})$ around the prior estimate $\mathbf{z}^{f}$,*

$$\mathbf{g}(\mathbf{z}) \approx \mathbf{g}(\mathbf{z}^{f}) + \mathbf{G}(\mathbf{z} - \mathbf{z}^{f}), \tag{6.4}$$

and approximate the gradient in Eq. (6.1) with the gradient evaluated at the prior estimate

$$\nabla_{\mathbf{z}}\mathbf{g}(\mathbf{z}) \approx \mathbf{G}^{T}, \tag{6.5}$$

where we have defined

$$\mathbf{G}^{T} = \nabla_{\mathbf{z}}\mathbf{g}(\mathbf{z})\big|_{\mathbf{z}=\mathbf{z}^{f}}. \tag{6.6}$$

Here, $\mathbf{G}$ is the tangent-linear operator of $\mathbf{g}(\mathbf{z})$ evaluated at $\mathbf{z}^{f}$, and $\mathbf{G}^{T}$ is its adjoint. Note that the Eq. (6.6) implies the following

$$\mathbf{M}_k^{T} = \nabla_{\mathbf{z}}\mathbf{m}(\mathbf{z})\big|_{\mathbf{z}=\mathbf{z}_k} \quad and \quad \mathbf{H}^{T} = \nabla_{\mathbf{m}(\mathbf{z})}\mathbf{h}(\mathbf{m}(\mathbf{z}))\big|_{\mathbf{z}=\mathbf{z}_k}. \tag{6.7}$$

The linearization in Eq. (6.4) and approximation in Eq. (6.5) allow us to rewrite Eq. (3.11) in terms of $\mathbf{g}(\mathbf{z}^{f})$ and $\mathbf{G}$, with an explicit solution

$$\mathbf{z}^{a} = \mathbf{z}^{f} + \left(\mathbf{C}_{zz}^{-1} + \mathbf{G}^{T}\mathbf{C}_{dd}^{-1}\mathbf{G}\right)^{-1}\mathbf{G}^{T}\mathbf{C}_{dd}^{-1}\left(\mathbf{d} - \mathbf{g}(\mathbf{z}^{f})\right). \tag{6.8}$$

We can write this equation in an alternative form by using the corollaries

---

Woodbury corollaries

$$\left(\mathbf{C}^{-1} + \mathbf{G}^{T}\mathbf{D}^{-1}\mathbf{G}\right)^{-1} = \mathbf{C} - \mathbf{C}\mathbf{G}^{T}(\mathbf{G}\mathbf{C}\mathbf{G}^{T} + \mathbf{D})^{-1}\mathbf{G}\mathbf{C}, \tag{6.9}$$

$$\left(\mathbf{G}^{T}\mathbf{D}^{-1}\mathbf{G} + \mathbf{C}^{-1}\right)^{-1}\mathbf{G}^{T}\mathbf{D}^{-1} = \mathbf{C}\mathbf{G}^{T}(\mathbf{G}\mathbf{C}\mathbf{G}^{T} + \mathbf{D})^{-1}, \tag{6.10}$$

---

which we derive from the Woodbury identity. We then obtain

$$\mathbf{z}^{a} = \mathbf{z}^{f} + \mathbf{C}_{zz}\mathbf{G}^{T}\left(\mathbf{G}\mathbf{C}_{zz}\mathbf{G}^{T} + \mathbf{C}_{dd}\right)^{-1}\left(\mathbf{d} - \mathbf{g}(\mathbf{z}^{f})\right), \qquad (6.11)$$

where we solve for the update in measurement space. Due to Approx. 5, the Eqs. (6.8) and (6.11) are only valid for small updates.

Using Eq. (6.11), we can compute an approximate update of the state vector $\mathbf{z}$ from measurements distributed over the assimilation window. Thus, the method solves a similar problem to the SC-4DVar discussed in Chap. 4 without using iterations.

Interestingly, from Eq. (6.3), the products $\mathbf{G}\mathbf{C}_{zz}$ and $\mathbf{G}\mathbf{C}_{zz}\mathbf{G}^{T}$ include a forward propagation of the background-error-covariance matrix $\mathbf{C}_{zz}$ leading to a covariance matrix for the model state over the whole data assimilation window. When we measure the resulting covariances, we obtain the covariances between the predicted measurements and the state vector, i.e., $\mathbf{C}_{yz} = \mathbf{G}\mathbf{C}_{zz}$ and $\mathbf{C}_{yy} = \mathbf{G}\mathbf{C}_{zz}\mathbf{G}^{T}$. With this, we can write Eq. (6.11) as

$$\mathbf{z}^{a} = \mathbf{z}^{f} + \mathbf{C}_{zy}\left(\mathbf{C}_{yy} + \mathbf{C}_{dd}\right)^{-1}\left(\mathbf{d} - \mathbf{g}(\mathbf{z}^{f})\right). \qquad (6.12)$$

In a prediction system, we would like to initialize the prediction for the next assimilation window. A question is whether it is possible to update the solution at the end of the assimilation window. Below, we will see that this is possible with the Kalman filter that provides an update and optimal solution at the end of the assimilation window by sequentially assimilating the measurements while evolving the model solution and its error statistics forward in time.

Let's revert to Eq. (6.11) and multiply the equation with $\mathcal{M}$ defined in Eq. (6.3) to obtain

$$\mathcal{M}\mathbf{z}^{a} = \mathcal{M}\mathbf{z}^{f} + \mathcal{M}\mathbf{C}_{zz}\mathcal{M}^{T}\mathbf{H}^{T}\left(\mathbf{G}\mathbf{C}_{zz}\mathbf{G}^{T} + \mathbf{C}_{dd}\right)^{-1}\left(\mathbf{d} - \mathbf{g}(\mathbf{z}^{f})\right). \qquad (6.13)$$

We can write this equation as

$$\begin{pmatrix} \mathbf{z}^{a} \\ \mathbf{x}_{1}^{a} \\ \vdots \\ \mathbf{x}_{K}^{a} \end{pmatrix} = \begin{pmatrix} \mathbf{z}^{f} \\ \mathbf{x}_{1}^{f} \\ \vdots \\ \mathbf{x}_{K}^{f} \end{pmatrix} + \begin{pmatrix} \mathbf{C}_{zz} & \cdots & \mathbf{C}_{zx_{K}} \\ \mathbf{C}_{x_{1}z} & \cdots & \mathbf{C}_{x_{1}x_{K}} \\ \vdots & \ddots & \vdots \\ \mathbf{C}_{x_{K}z} & \cdots & \mathbf{C}_{x_{K}x_{K}} \end{pmatrix} \mathbf{H}^{T}\left(\mathbf{G}\mathbf{C}_{zz}\mathbf{G}^{T} + \mathbf{C}_{dd}\right)^{-1}\left(\mathbf{d} - \mathbf{g}(\mathbf{z}^{f})\right). \quad (6.14)$$

It is clear that this formulation gives a smoother update of the model solution over the whole assimilation window, processing all the measurements in one go, as discussed in Sect. 2.4.1. If we are only interested in the solution at the time t_{K}, we can compute the following

$$\mathbf{x}_{K}^{a} = \mathbf{x}_{K}^{f} + \left(\mathbf{C}_{x_{K}z} \cdots \mathbf{C}_{x_{K}x_{K}}\right)\mathbf{H}^{T}\left(\mathbf{G}\mathbf{C}_{zz}\mathbf{G}^{T} + \mathbf{C}_{dd}\right)^{-1}\left(\mathbf{d} - \mathbf{g}(\mathbf{z}^{f})\right) \qquad (6.15)$$

$$= \mathbf{x}_{K}^{f} + \mathbf{C}_{x_{K}y}\left(\mathbf{C}_{yy} + \mathbf{C}_{dd}\right)^{-1}\left(\mathbf{d} - \mathbf{g}(\mathbf{z}^{f})\right). \qquad (6.16)$$

As for the case where we updated the initial model state of the assimilation window, we now use the covariance matrix $\mathbf{C}_{x_{K}y}$ to update the final model state of the assimilation window. Using Eq. (6.14), we can update the model-state vector at any time within the assimilation window.

In Eq. (6.14), we must integrate the model to predict $\mathbf{x}_K^f$ and the covariance matrix through $\mathcal{M}\mathbf{C}_{zz}\mathcal{M}^T$, where the equation to update $\mathcal{M}\mathbf{C}_{zz}\mathcal{M}^T$ is similar to the error covariance equation from Eq. (2.28) without explicit model errors.

From the above, we learn that it is possible to update the model state at a particular time using measurements distributed in time by exploiting the time correlations. Sakov et al. (2010) discussed this "asynchronous" data assimilation in the ensemble Kalman filter. They showed how to assimilate batches of multiple measurements distributed in time to avoid stopping and restarting the model integration too frequently.

The above computations are not convenient, but the approach becomes practical with the ensemble data-assimilation methods. Finally, note that in this linear case, without model errors, updating $\mathbf{x}_0$ and propagating this solution forward to time t_K to find $\mathbf{x}_K$ gives the same result for $\mathbf{x}_K$ as if we first run the model to obtain a forecast of $\mathbf{x}_K$ and then compute its update.

6.2   3DVar

3DVar used to be a popular approach to minimize the cost function from Eq. (3.9), assuming that the prior and the measurements are both available at the initial time of the assimilation window. This "time-independence" implies that the possibly nonlinear function $\mathbf{g}(\mathbf{z})$ represents only the measurement operator and not the model operator. In this case, the state vector $\mathbf{z}$ includes the model state at the initial time of the assimilation window and may contain model parameters. We then start from the cost function

---

3DVar costfunction

$$\mathcal{J}(\mathbf{z}) = \frac{1}{2}\left(\mathbf{z} - \mathbf{z}^f\right)^T \mathbf{C}_{zz}^{-1}\left(\mathbf{z} - \mathbf{z}^f\right) + \frac{1}{2}\left(\mathbf{h}(\mathbf{z}) - \mathbf{d}\right)^T \mathbf{C}_{dd}^{-1}\left(\mathbf{h}(\mathbf{z}) - \mathbf{d}\right). \qquad (6.17)$$

The 3DVar method refers specifically to a sequential data-assimilation approach where we use a constant-in-time background or prior error covariance $\mathbf{C}_{zz}$ for each subsequent update step. Thus, the method does not propagate error statistics from one update time till the next, and there is no updating of the analysis error covariance. This 3DVar update scheme solves a Gauss–Newton iteration like in Eqs. (3.14) or (3.17). Thus, 3DVar is a computationally efficient, although approximate method.

The gradient of the cost function is still the one in Eq. (3.10) but with $\mathbf{g}(\mathbf{z})$ replaced by $\mathbf{h}(\mathbf{z})$, i.e.,

$$\nabla_{\mathbf{z}}\mathcal{J}(\mathbf{z}) = \mathbf{C}_{zz}^{-1}\left(\mathbf{z} - \mathbf{z}^f\right) + \mathbf{H}^T \mathbf{C}_{dd}^{-1}\left(\mathbf{h}(\mathbf{z}) - \mathbf{d}\right), \qquad (6.18)$$

where we have used

$$\mathbf{H}^T = \nabla_{\mathbf{z}}\mathbf{h}(\mathbf{z})\big|_{\mathbf{z}=\mathbf{z}}. \qquad (6.19)$$

No explicit solution solves Eq. (6.18) being equal to zero since the gradient includes the nonlinear measurement operator $\mathbf{h}(\mathbf{z})$. Thus, to compute the 3DVar solution, we must use an iterative solver. Another reason for using an iterative approach is that we avoid forming the explicit model state covariance matrix and invert it, as in the 4DVar methods.

Typically, we initialize a Gauss–Newton iteration with the prior estimate

$$\mathbf{z}^0 = \mathbf{z}^f, \tag{6.20}$$

and iterate

$$\mathbf{z}^{i+1} = \mathbf{z}^i - \gamma^i \mathbf{B}^i \nabla_{\mathbf{z}} \mathcal{J}(\mathbf{z}^i), \tag{6.21}$$

until convergence. In this expression, $\mathbf{B}^i$ is the inverse of the Hessian given by

$$\mathbf{C}_{zz}^{-1} + \mathbf{H}^{i\,\mathrm{T}} \mathbf{C}_{dd}^{-1} \mathbf{H}^i. \tag{6.22}$$

If we use the gradient from Eq. (6.18), we can write the iteration in Eq. (6.21) as

$$\mathbf{z}^{i+1} = \mathbf{z}^i - \gamma^i \left(\mathbf{C}_{zz}^{-1} + \mathbf{H}^{i\,\mathrm{T}} \mathbf{C}_{dd}^{-1} \mathbf{H}^i \right)^{-1} \left(\mathbf{C}_{zz}^{-1} (\mathbf{z}^i - \mathbf{z}^f) + \mathbf{H}^{i\,\mathrm{T}} \mathbf{C}_{dd}^{-1} \left(\mathbf{h}(\mathbf{z}^i) - \mathbf{d} \right) \right) \tag{6.23}$$

$$= \mathbf{z}^i - \gamma^i (\mathbf{z}^i - \mathbf{z}^f)$$
$$+ \gamma^i \mathbf{C}_{zz} \mathbf{H}^{\mathrm{T}} \left(\mathbf{H} \mathbf{C}_{zz} \mathbf{H}^{\mathrm{T}} + \mathbf{C}_{dd} \right)^{-1} \left(\mathbf{H}(\mathbf{z}^i - \mathbf{z}^f) - \left(\mathbf{h}(\mathbf{z}^i) - \mathbf{d} \right) \right), \tag{6.24}$$

where we again used the corollaries from Eqs. (6.9) and (6.10).

Instead of forming the Hessian and inverting, it is common to introduce the control-variable transform as in strong-constraint 4Dvar. It is then possible to apply iterative methods like conjugate-gradient to solve the linearized problem.

An advantage of 3DVar is that it minimizes the cost function with a nonlinear measurement operator. The method is highly efficient as it does not update or evolve error statistics in time. Still, the approximation of a constant-in-time background-error-covariance matrix can be an essential drawback of 3DVar. Furthermore, while 3DVar solves for the update with a nonlinear measurement functional, we can compute the update even more efficiently in the case with a linear measurement functional using the Kalman-filter update equations. With the assumptions on linearity, this approach avoids iterations and leads to the Optimal Interpolation method in the case of stationary error statistics.

6.3   Kalman Filter

Generally, it is only possible to write a closed-form solution of the Eq. (3.11) in the trivial case when $\mathbf{g}(\mathbf{z})$ is linear. In the case with a linear measurement functional, and when the measurements are available at the initial time of the assimilation window, Eq. (3.11) reduces to Eq. (6.17). With a linear measurement operator, $\mathbf{h}(\mathbf{z})$ becomes $\mathbf{H}\mathbf{z}$, and we write the cost function in Eq. (6.17) as

$$\mathcal{J}(\mathbf{z}) = \frac{1}{2} (\mathbf{z} - \mathbf{z}^f)^{\mathrm{T}} \mathbf{C}_{zz}^{-1} (\mathbf{z} - \mathbf{z}^f) + \frac{1}{2} (\mathbf{H}\mathbf{z} - \mathbf{d})^{\mathrm{T}} \mathbf{C}_{dd}^{-1} (\mathbf{H}\mathbf{z} - \mathbf{d}). \tag{6.25}$$

By setting the gradient of the cost function to zero,

$$\mathbf{C}_{zz}^{-1}\left(\mathbf{z}^{a} - \mathbf{z}^{f}\right) + \mathbf{H}^{T}\,\mathbf{C}_{dd}^{-1}\left(\mathbf{H}\mathbf{z}^{a} - \mathbf{d}\right) = 0, \tag{6.26}$$

we find the Kalman filter update equation

$$\mathbf{z}^{a} = \mathbf{z}^{f} + \left(\mathbf{C}_{zz}^{-1} + \mathbf{H}^{T}\mathbf{C}_{dd}^{-1}\mathbf{H}\right)^{-1}\mathbf{H}^{T}\mathbf{C}_{dd}^{-1}\left(\mathbf{d} - \mathbf{H}\mathbf{z}^{f}\right), \tag{6.27}$$

We solve the analysis $\mathbf{z}^{a}$ in the state space because we define the matrices we invert in the state space. By using the matrix identity from Eq. (6.10), we can rewrite Eq. (6.27) to obtain the standard form of the Kalman filter update equation

---

The Kalman filter state update

$$\mathbf{z}^{a} = \mathbf{z}^{f} + \mathbf{C}_{zz}\mathbf{H}^{T}\left(\mathbf{H}\mathbf{C}_{zz}\mathbf{H}^{T} + \mathbf{C}_{dd}\right)^{-1}\left(\mathbf{d} - \mathbf{H}\mathbf{z}^{f}\right), \tag{6.28}$$

---

which solves for the solution in the measurement space just like the formulation in Eq. (6.11). Indeed, the matrix we have to invert is defined in the measurement space.

We can find the Hessian of the cost function in Eq. (6.25) by taking the second derivative, leading to

$$\nabla_{z}\nabla_{z}\mathcal{J}(\mathbf{z}) = \mathbf{C}_{zz}^{-1} + \mathbf{H}\mathbf{C}_{dd}^{-1}\mathbf{H}^{T}. \tag{6.29}$$

Note that the Hessian is not dependent on the state $\mathbf{z}$ but only on the prior and the measurement covariances. We know that the posterior is Gaussian, with covariance matrix $\mathbf{C}_{zz}^{a}$. Hence, we can write the cost function as

$$\mathcal{J}(\mathbf{z}) = \frac{1}{2}\left(\mathbf{z} - \mathbf{z}^{a}\right)^{T}\left(\mathbf{C}_{zz}^{a}\right)^{-1}\left(\mathbf{z} - \mathbf{z}^{a}\right) + constant \tag{6.30}$$

in which the constant depends on the measurement $\mathbf{d}$, but not on $\mathbf{z}$. Taking the second derivative of this version of the cost function gives

$$\nabla_{z}\nabla_{z}\mathcal{J}(\mathbf{z}^{a}) = \left(\mathbf{C}_{zz}^{a}\right)^{-1}. \tag{6.31}$$

Since the two expressions for the Hessian must be the same, we find for the posterior covariance

$$\left(\mathbf{C}_{zz}^{a}\right)^{-1} = \mathbf{C}_{zz}^{-1} + \mathbf{H}\mathbf{C}_{dd}^{-1}\mathbf{H}^{T}. \tag{6.32}$$

With this expression, we have proven that in the linear case, the inverse of the Hessian is the posterior error-covariance matrix. Eq. (6.32) is the equation for computing the error-covariance matrix in the state space, and we can rewrite it using the matrix identity (6.9) to find

---

The Kalman filter error-covariance update

$$\mathbf{C}_{zz^{a}} = \mathbf{C}_{zz} - \mathbf{C}_{zz}\mathbf{H}^{T}\left(\mathbf{H}\mathbf{C}_{zz}\mathbf{H}^{T} + \mathbf{C}_{dd}\right)^{-1}\mathbf{H}\mathbf{C}_{zz}. \tag{6.33}$$

In contrast to using a stationary background-error-covariance matrix in 3DVar, the Kalman filter updates and evolves the error statistics in time, using Eq. (2.28). The standard form for the Kalman filter is a recursion over measurement times where we evolve the state vector and its covariance from one measurement time till the next by solving the following model and error covariance equations, starting from the update $\mathbf{z}_k = \mathbf{z}^a$ and $\mathbf{C}_{zz,k} = \mathbf{C}_{zz^a}$ at the time t_k

$$\mathbf{z}_{k+1} = \mathbf{M}\mathbf{z}_k, \tag{6.34}$$

$$\mathbf{C}_{zz,k+1} = \mathbf{M}\mathbf{C}_{zz,k}\mathbf{M}^{\mathrm{T}} + \mathbf{C}_{qq}. \tag{6.35}$$

We integrate these equations until the next time t_k when we update the model with new measurements. We then set $\mathbf{z}^f = \mathbf{z}_k$ and $\mathbf{C}_{zz} = \mathbf{C}_{zz,k}$, and we compute the update from Eqs. (6.28) and (6.33), before we continue the integration of Eqs. (6.34) and (6.35).

Thus, we define the assimilation windows in the KF to cover the time intervals between two consecutive measurement times. We integrate the model solution and the error covariance matrix from the start of an assimilation window till the next. We then update the predicted model state vector and its error covariance matrix at the initial time of the next assimilation window. We repeat this recursion as we progress from one assimilation window to the next.

In the case of a linear measurement operator and Gaussian priors, Eqs. (6.28) and (6.33) provide the variance minimizing solution of

$$f(\mathbf{z}|\mathbf{d}) \propto \exp\left(-\frac{1}{2}\mathcal{J}(\mathbf{z})\right), \tag{6.36}$$

with cost function $\mathcal{J}$ defined in Eq. (6.25). We note that the variance-minimizing solution is equal to the MAP solution for a purely Gaussian problem. Thus, the Eqs. (6.28) and (6.33) exactly represent the posterior pdf for the Gauss-linear case as described in the cost function in Eq. (3.9).

One of the main issues with the KF is the storage of the vast error-covariance matrix $\mathbf{C}_{zz}$ and the computational cost of its evolution in time. For an example of using the KF and its properties, we refer to Chaps. 12 and 13.

We observe that we can also write the KF update in Eq. (6.28) using the notation

$$\mathbf{z}^a = \mathbf{z}^f + \mathbf{K}\left(\mathbf{d} - \mathbf{h}\left(\mathbf{z}^f\right)\right), \tag{6.37}$$

$$\mathbf{C}_{zz^a} = \left(\mathbf{I} - \mathbf{K}\mathbf{H}\right)\mathbf{C}_{zz}, \tag{6.38}$$

where $\mathbf{K} \in \mathfrak{R}^{n \times m}$ defines the Kalman gain matrix of the Kalman filter (Kalman, 1960),

$$\mathbf{K} = \mathbf{C}_{zz}\mathbf{H}^{\mathrm{T}}\left(\mathbf{H}\mathbf{C}_{zz}\mathbf{H}^{\mathrm{T}} + \mathbf{C}_{dd}\right)^{-1}, \tag{6.39}$$

and Eqs. (6.37)–(6.39) represent the update step of the Kalman filter.

This section showed that the solution of Eq. (3.11) is defined in the state space, i.e., we solve for $\mathbf{z}^a$. However, we have transformed the update computation to the measurement space using the Woodbury identity, identifiable by the matrix that we

must invert in the Kalman filter. Bennett (1992) showed that it is possible to write
the exact solution of Eq. (6.28) as

$$\mathbf{z}^a = \mathbf{z}^f + \mathbf{C}_{zy}\mathbf{b}.\tag{6.40}$$

which is a linear combination of "representer functions"

$$\mathbf{C}_{zy} = \mathbf{C}_{zz}\mathbf{H}^T,\tag{6.41}$$

with coefficients $\mathbf{b}$ found by solving the linear system

$$\left(\mathbf{H}\mathbf{C}_{zz}\mathbf{H}^T + \mathbf{C}_{dd}\right)\mathbf{b} = \mathbf{d} - \mathbf{H}\mathbf{z}^f.\tag{6.42}$$

 The update to $\mathbf{z}^f$ resides in an m-dimensional space in this formulation, hence the
notation measurement space. The notation state space makes sense as the state vector
belongs to it. On the other hand, the notation measurement space refers to the fact
that we first find the solution in the measurement space, and then transform it to the
state space via the $\mathbf{C}_{zy}$ matrix. We are still computing the solution in the state space,
but the update is a linear combination of m representer functions $\mathbf{C}_{zy}$. Hence, as we
calculate the update to $\mathbf{z}^f$ in the space spanned by the m representer functions, we
could also have used a name like "representer space." The critical point is that we
reduce the inverse calculations from an n-dimensional problem to an m-dimensional
one.

6.4 Optimal Interpolation

From the KF formulation, we obtain an even more straightforward sequential updating algorithm named optimal interpolation (OI) by applying the approximation
of time-invariant error statistics. In OI, we only solve for the model prediction in
Eq. (6.34) and update the model state according to Eq. (6.28), with $\mathbf{C}_{zz}$ being a
constant-in-time prior error-covariance matrix. In the linear case, optimal interpolation is equivalent to the 3DVar method. However, the iterative solution method used
in 3DVar allows for solving the update with a nonlinear measurement functional.
Avoiding a matrix inversion can be much more efficient, even in the linear case.

6.5 Extended Kalman Filter

The extended Kalman filter (EKF) allows applying the KF with a nonlinear model
and measurement operator. To derive the equations for the update step, we again
start with the 3DVar cost function in Eq. (6.17) and its gradient Eq. (6.18). To find
an explicit but approximate solution $\mathbf{z}$ of the gradient in Eq. (6.18) equal to zero, we
use the Approx. 5, which allows us to write

$$\mathbf{C}_{zz}^{-1}\left(\mathbf{z}^a - \mathbf{z}^f\right) + \mathbf{H}^T\mathbf{C}_{dd}^{-1}\left(\mathbf{h}(\mathbf{z}^f) + \mathbf{H}(\mathbf{z}^a - \mathbf{z}^f) - \mathbf{d}\right).\tag{6.43}$$

Like for the KF, we find an explicit solution of this equation as

$$\mathbf{z}^{a} = \mathbf{z}^{f} + \mathbf{C}_{zz}\mathbf{H}^{T}\left(\mathbf{H}\mathbf{C}_{zz}\mathbf{H}^{T} + \mathbf{C}_{dd}\right)^{-1}\left(\mathbf{d} - \mathbf{h}\left(\mathbf{z}^{f}\right)\right). \tag{6.44}$$

The update equation becomes identical to the Eq. (6.33) used in the Kalman filter but now uses $\mathbf{h}$ instead of $\mathbf{H}$ when comparing the measurements to the model prediction. To derive an equation for the error covariance evolution, we need to linearize the model equations. By introducing the linearization from Eq. (6.7) in Approx. 5, we can compute the time evolution of the model state and its error covariance from

$$\mathbf{z}_{k+1} = \mathbf{m}\left(\mathbf{z}_{k}\right) \tag{6.45}$$

$$\mathbf{C}_{zz,k+1} = \mathbf{M}_{k}\mathbf{C}_{zz,k}\mathbf{M}_{k}^{T} + \mathbf{C}_{qq}, \tag{6.46}$$

where Eq. (6.46) was derived in Sect. 2.3.2.

Note that the EKF applies linear versions of the model and measurement operators. The state vector $\mathbf{z}^{f}$ contains the predicted model state at an update time originating from the nonlinear model Eq. (2.5). However, an approximate linearized equation (2.28) describes the time evolution of the state-error-covariance matrix. Evensen (1992 found that using a linearized error-covariance equation led to linear instabilities, which would cause the predicted error covariance to blow up for many nonlinear models with unstable dynamics such as ocean and atmospheric models. We will discuss the EKF further in the example in Chap. 12.

Randomized-Maximum-Likelihood Sampling

7

In the following, we derive some methods for sampling the posterior conditional pdf in Eq. (3.8). We aim to estimate the full pdf, not only finding its maximum. We will, in this chapter, use an approach named randomized maximum likelihood (RML) sampling. Note that the name is not precise as the method attempts to sample the posterior pdf and not just the likelihood. However, we will continue using the name RML when we refer to the technique. RML provides a highly efficient approach for approximate sampling of the posterior pdf and lays the ground for developing many popular ensemble methods.

7.1 RML Sampling

To introduce randomized-maximum-likelihood sampling, let's define an ensemble of cost functions where the prior vectors $\mathbf{z}_j^f$ are samples from the Gaussian distribution in Eq. (3.5), and we introduce the perturbed measurements $\mathbf{d}_j = \mathbf{d} + \boldsymbol{\epsilon}_j$ where the perturbations $\boldsymbol{\epsilon}_j$ are samples from (3.6),

---

Ensemble of cost functions

$$\mathcal{J}(\mathbf{z}_j) = \frac{1}{2}\left(\mathbf{z}_j - \mathbf{z}_j^f\right)^{\mathrm{T}} \mathbf{C}_{zz}^{-1}\left(\mathbf{z}_j - \mathbf{z}_j^f\right) + \frac{1}{2}\left(\mathbf{g}(\mathbf{z}_j) - \mathbf{d}_j\right)^{\mathrm{T}} \mathbf{C}_{dd}^{-1}\left(\mathbf{g}(\mathbf{z}_j) - \mathbf{d}_j\right), \quad (7.1)$$

---

as proposed by Kitanidis, (1995) and Oliver et al., (1996). These cost functions are independent of each other and differ from the cost function (3.9) by the introduction of the random samples $\mathbf{z}_j^f \sim \mathcal{N}(\mathbf{z}^f, \mathbf{C}_{zz})$ and $\mathbf{d}_j \sim \mathcal{N}(\mathbf{d}, \mathbf{C}_{dd})$.

One might ask why we need to perturb the measurements when Bayes theorem tells us that we are already given the measurements in the data-assimilation

© The Author(s) 2022
G. Evensen et al., *Data Assimilation Fundamentals*,
Springer Textbooks in Earth Sciences, Geography and Environment,
https://doi.org/10.1007/978-3-030-96709-3_7

problem. Indeed, Van Leeuwen (2020) contains a detailed discussion on why it is more consistent to perturbing the predicted measurements $\mathbf{g}(\mathbf{z}_j)$ with a draw from the measurement error pdf. There is no practical advantage for either choice. The reason is that the cost function in Eq. (7.1) only contains the difference between the predicted and actual measurements, and the Gaussian is symmetric in its arguments. In this chapter, we will use the conventional "perturbed measurements" formalism.

Approximation 6 (RML sampling) *In the weakly nonlinear case, we can approximately sample the posterior pdf in Eq. (3.8) by minimizing the ensemble of cost functions defined by Eq. (7.1).* $\square$

In the Gauss-linear case, the minimizing solutions of these cost functions precisely sample the posterior conditional pdf in Eq. (3.8). Furthermore, with an infinite number of samples, the sample mean and covariance will converge to the KF solution given by Eqs. (6.28) and (6.33). When we introduce nonlinearity into the problem, the samples will deviate from the pdf in Eq. (3.8). But in many cases with only weak nonlinearity, this approximation is acceptable. The fun fact is that nobody knows precisely which distribution the method samples in the nonlinear case. Note also that we can minimize each of the cost functions independently of the others using the Gauss–Newton method described in Chap. 3.

Similarly to Eq. (3.11), we now have an ensemble of gradients that we set to zero to minimize the ensemble of cost functions in Eq. (7.1),

---

Ensemble of gradients set to zero

$$\mathbf{C}_{zz}^{-1}\left(\mathbf{z}_j - \mathbf{z}_j^f\right) + \nabla_{\mathbf{z}}\mathbf{g}(\mathbf{z}_j)\mathbf{C}_{dd}^{-1}\left(\mathbf{g}(\mathbf{z}_j) - \mathbf{d}_j\right) = 0. \tag{7.2}$$

7.2 Approximate EKF Sampling

The simplest way to solve Eq. (7.2) for an ensemble of realizations is to use the Kalman filter update Eq. (6.44) to solve for each sample, $j = 1, \ldots, N_{ens}$,

---

Ensemble of Kalman-filter updates

$$\mathbf{z}_j^a = \mathbf{z}_j^f + \mathbf{C}_{zz}\mathbf{G}_j^T\left(\mathbf{G}_j\mathbf{C}_{zz}\mathbf{G}_j^T + \mathbf{C}_{dd}\right)^{-1}\left(\mathbf{d}_j - \mathbf{g}(\mathbf{z}_j^f)\right). \tag{7.3}$$

However, as we noted in the previous chapter, these equations are only valid in the linear case or for modest updates in the nonlinear case.

7.3 Approximate Gauss–Newton Sampling

As an alternative to the EKF solution from Sect. 6.5, we can minimize the cost function in Eq. (7.1) without introducing the Approx. 5. We do this by using the Gauss–Newton method as in Sect. 3.4 for each of the cost functions in the ensemble. Taking the derivative of Eq. (3.10) while neglecting terms including second derivatives, we obtain an approximation to the Hessian

$$\nabla_{\mathbf{z}}\nabla_{\mathbf{z}}\mathcal{J}(\mathbf{z}_j) \approx \mathbf{C}_{zz}^{-1} + \nabla_{\mathbf{z}}\mathbf{g}(\mathbf{z}_j)\mathbf{C}_{dd}^{-1}\left(\nabla_{\mathbf{z}}\mathbf{g}(\mathbf{z}_j)\right)^{\mathrm{T}}. \tag{7.4}$$

We can then write a Gauss–Newton iteration for $\mathbf{z}$ as

---

Ensemble of GN iterations

$$\mathbf{z}_j^{i+1} = \mathbf{z}_j^i - \gamma \left(\mathbf{C}_{zz}^{-1} + \mathbf{G}_j^{i\,\mathrm{T}} \mathbf{C}_{dd}^{-1} \mathbf{G}_j^i \right)^{-1} \left(\mathbf{C}_{zz}^{-1} \left(\mathbf{z}_j^i - \mathbf{z}_j^f \right) + \mathbf{G}_j^{i\,\mathrm{T}} \mathbf{C}_{dd}^{-1} \left(\mathbf{g}\left(\mathbf{z}_j^i\right) - \mathbf{d}_j \right) \right),$$
$$\tag{7.5}$$

---

where we have defined the gradient of the observation operator at iteration i and for ensemble member j as

$$\mathbf{G}_j^{i\,\mathrm{T}} = \nabla_{\mathbf{z}}\mathbf{g}\left(\mathbf{z}_j^i\right). \tag{7.6}$$

In this formulation, each realization uses the tangent-linear model $\mathbf{G}_j^i$ evaluated at the solution for realization j at iteration i. Thus, each realization has a model sensitivity that is independent of the other realizations. This approach and any other method that minimizes the cost functions in Eq. (7.1) will correctly sample the posterior distribution in the Gauss-linear case. Still, for a posterior non-Gaussian distribution, Approx. 6 applies. Thus, we can use any of the methods discussed in Chaps. 3, 4, and 5 to solve for the minimizing solution of each cost-function realization.

7.4 Least-Squares Best-Fit Model Sensitivity

There are two aspects of the solutions defined in Eqs. (7.3) and (7.5) that require our attention. First, we assume we know the tangent-linear model $\mathbf{G}_j^i$ and its adjoint, $\mathbf{G}_j^{i\,\mathrm{T}}$, which is not always the case. The other aspect relates to the storage and inversion of $\mathbf{C}_{zz}$, a huge matrix.

In cases when we do not have access to a tangent-linear model or the adjoint operator, we can use a statistical representation of the model sensitivity. Rather than computing different tangent linear operators $\mathbf{G}_j^i$ for each sample, we represent them by a statistical least-squares best-fit model sensitivity $\mathbf{G}^i$ common for all realizations (Chen & Oliver, 2013; Evensen, 2019; Reynolds et al., 2006), and we introduce the following approximation.

Approximation 7 (Best-fit ensemble-averaged model sensitivity) *Interpret* $\mathbf{G}_j$
in Eq. (7.3) *and* $\mathbf{G}_j^i$ *in Eq.* (7.5) *as the sensitivity matrix in linear regression and*
represent them using the definition

$$\mathbf{G}_j \approx \mathbf{G} \triangleq \mathbf{C}_{yz}\mathbf{C}_{zz}^{-1}. \tag{7.7}$$

Note that we have dropped the superscript j *for the realizations. Hence, we ap-*
proximate the individual model sensitivities with a common averaged sensitivity
used for all realizations.

A consequence of this approximation is that we slightly alter the gradient in Eq. (7.2)
and thus also the minimizing solution that the Kalman filter updates or the Gauss–
Newton iterations would provide.

By introducing the averaged model-sensitivity from Eq. (7.7), we can rewrite the
Gauss–Newton iteration in Eq. (7.5) as

$$\mathbf{z}_j^{i+1} = \mathbf{z}_j^i - \gamma\left(\mathbf{C}_{zz}^{-1} + \mathbf{G}^{i\mathrm{T}}\mathbf{C}_{dd}^{-1}\mathbf{G}^i\right)^{-1}\left(\mathbf{C}_{zz}^{-1}\left(\mathbf{z}_j^i - \mathbf{z}_j^f\right) + \mathbf{G}^{i\mathrm{T}}\mathbf{C}_{dd}^{-1}\left(\mathbf{g}(\mathbf{z}_i) - \mathbf{d}_j\right)\right) \tag{7.8}$$

$$= \mathbf{z}_j^i - \gamma\left(\mathbf{z}_j^i - \mathbf{z}_j^f\right)$$
$$+ \gamma\mathbf{C}_{zz}\mathbf{G}^{i\mathrm{T}}\left(\mathbf{G}^i\mathbf{C}_{zz}\mathbf{G}^{i\mathrm{T}} + \mathbf{C}_{dd}\right)^{-1}\left(\mathbf{G}^i\left(\mathbf{z}_j^i - \mathbf{z}_j^f\right) - \left(\mathbf{g}(\mathbf{z}_j^i) - \mathbf{d}_j\right)\right), \tag{7.9}$$

where we have used the corollaries from Eqs. (6.9) and (6.10).

A rather tricky issue with Eq. (7.9) is the appearance of products between the
averaged model sensitivity $\mathbf{G}^i$ evaluated at iteration i with the prior covariance matrix
$\mathbf{C}_{zz}$. Chen and Oliver (2013) provided an alternative approach by evaluating the state
covariance in the Hessian at the current iterate. This modification does not impact
the final solution, but it alters the update step. They introduced various strategies for
solving Eqs. (7.8) and (7.9) using ensemble representations for the state covariances.
The next chapter will present a recent and efficient algorithm that searches for the
solution in the ensemble subspace.

Recall that $\mathbf{y} = \mathbf{g}(\mathbf{z})$ is the model equivalent of the observed state and $\mathbf{C}_{yz}$ is
the covariance between the state vector $\mathbf{z}$ and the predicted measurements $\mathbf{y}$. The
operator $\mathbf{G}$, defined in Eq. (7.7), is the linear regression between $\mathbf{y}$ and $\mathbf{z}$, and we
have

$$\mathbf{G}\mathbf{C}_{zz} = \mathbf{C}_{yz}, \tag{7.10}$$

and

$$\mathbf{G}\mathbf{C}_{zz}\mathbf{G}^{\mathrm{T}} = \mathbf{C}_{yz}\mathbf{C}_{zz}^{-1}\mathbf{C}_{zy}. \tag{7.11}$$

We will use these expressions further in the following chapter. For now, we note
that we can use the EKF update Eq. (7.3) to formulate an ensemble of Kalman-filter
updates without using the tangent-linear operator, as

$$\mathbf{z}_j = \mathbf{z}_j^f + \mathbf{C}_{zy}\left(\mathbf{C}_{yz}\mathbf{C}_{zz}^{-1}\mathbf{C}_{zy} + \mathbf{C}_{dd}\right)^{-1}\left(\mathbf{d}_j - \mathbf{g}(\mathbf{z}_j^f)\right). \tag{7.12}$$

It is common to replace the term $\mathbf{G}\mathbf{C}_{zz}\mathbf{G}^{\mathrm{T}} = \mathbf{C}_{yz}\mathbf{C}_{zz}^{-1}\mathbf{C}_{zy}$ with $\mathbf{C}_{yy}$. However, most
data-assimilation practitioners are unaware that this replacement introduces another

approximation if $\mathbf{g}(\mathbf{z})$ is nonlinear. In the following chapter, we will come back to this issue when discussing a low-rank ensemble approximation of the prior covariance matrix that leads to efficient ensemble-data-assimilation methods.

Low-Rank Ensemble Methods

8

This chapter will introduce another approximation where we represent all state error covariances using a finite ensemble of the state. This approximation allows us to search for the solution in the ensemble subspace, leading to very efficient ensemble data-assimilation methods. The most well-known is the ensemble Kalman filter, but we also have newer advanced schemes like the ensemble-randomized-maximum-likelihood method. In this chapter, we introduce the ensemble approximation and derive these ensemble subspace methods. We also illustrate using a single algorithm, i.e., an ensemble-subspace RML formulation, to compute the update in several traditional ensemble methods.

8.1 Ensemble Approximation

To ease the computational aspects of the methods discussed in the previous chapter, let's introduce a new approximation

> **Approximation 8** (Ensemble approximation) *It is possible to approximately represent a covariance matrix by a low-rank ensemble of states with fewer realizations than the state dimension.* □

Ensemble data-assimilation methods (Evensen, 1994) use a finite ensemble of state vectors to approximate the prior error covariance matrix C_{zz}. It is easy to show that we restrict the data-assimilation estimate to the ensemble space when representing C_{zz} by an ensemble of state vectors. This approach significantly simplifies the computational problem. In the following, we will introduce the ensemble covariance matrices into the Kalman-filter update and Gauss–Newton methods described in Chaps. 3 and 6.

© The Author(s) 2022
G. Evensen et al., *Data Assimilation Fundamentals*,
Springer Textbooks in Earth Sciences, Geography and Environment,
https://doi.org/10.1007/978-3-030-96709-3_8

8.2 Definition of Ensemble Matrices

We start by defining the prior ensemble of N model realizations, $\mathbf{z}_j \in \Re^n$, stored in the ensemble matrix $\mathbf{Z} \in \Re^{n \times N}$,

$$\mathbf{Z} = \left(\mathbf{z}_1^f, \ \mathbf{z}_2^f, \ \dots, \ \mathbf{z}_N^f \right). \tag{8.1}$$

Furthermore, we define the projection $\mathbf{\Pi} \in \Re^{N \times N}$ as

$$\mathbf{\Pi} = \left(\mathbf{I} - \frac{1}{N} \mathbf{11}^T \right) / \sqrt{N-1}, \tag{8.2}$$

where $\mathbf{1} \in \Re^N$ is a vector with all elements equal to one and $\mathbf{I}_N$ is the N-dimensional identity matrix. If we multiply an ensemble matrix with the orthogonal projection $\mathbf{\Pi}$, this subtracts the mean from the ensemble and scales the result with $1/\sqrt{N-1}$.

We can then define the zero-mean and scaled ensemble-anomaly matrix as

$$\mathbf{A} = \mathbf{Z}\mathbf{\Pi}. \tag{8.3}$$

Thus, the ensemble covariance is

$$\overline{\mathbf{C}}_{zz} = \mathbf{A}\mathbf{A}^T, \tag{8.4}$$

where the "overbar" denotes that we have an *ensemble*-covariance matrix.

Correspondingly, we can define an ensemble of perturbed measurements, $\mathbf{D} \in \Re^{m \times N}$, when given the real measurement vector, $\mathbf{d} \in \Re^m$, as

$$\mathbf{D} = \mathbf{d}\mathbf{1}^T + \sqrt{N-1}\mathbf{E}, \tag{8.5}$$

where $\mathbf{E} \in \Re^{m \times N}$ is the centered measurement-perturbation matrix whose columns are sampled from $\mathcal{N}(0, \mathbf{C}_{dd})$ and divided by $\sqrt{N-1}$. Thus, we define the ensemble covariance matrix for the measurement perturbations as

$$\overline{\mathbf{C}}_{dd} = \mathbf{E}\mathbf{E}^T. \tag{8.6}$$

The ensemble algorithms derived below work both with a full-rank $\mathbf{C}_{dd}$ or the ensemble version represented by the perturbations in $\mathbf{E}$.

Finally, we define the ensemble of model-predicted measurements

$$\mathbf{\Upsilon} = \mathbf{g}(\mathbf{Z}), \tag{8.7}$$

with anomalies

$$\mathbf{Y} = \mathbf{\Upsilon}\mathbf{\Pi}, \tag{8.8}$$

where we have multiplied the model prediction by the projection $\mathbf{\Pi}$ to subtract the ensemble mean and dividing the resulting anomalies by $\sqrt{N-1}$.

8.3   Cost Function in the Ensemble Subspace

We now introduce the ensemble representation from Eq. (8.4) into the approximate EnKF sampling in Eq. (7.3) or the RML sampling in Eqs. (7.8) or (7.9). It is then easy to show that the updated samples are confined to the space spanned by the prior ensemble since the leftmost matrix in the gradient is the ensemble anomaly matrix.

Thus, we will search for the solution in the ensemble subspace spanned by the prior ensemble by assuming that an updated ensemble realization, $\mathbf{z}_j^{\mathrm{a}}$, is equal to the prior realization, $\mathbf{z}_j^{\mathrm{f}}$, plus a linear combination of the ensemble anomalies,

$$\mathbf{z}_j^{\mathrm{a}} = \mathbf{z}_j^{\mathrm{f}} + \mathbf{A}\mathbf{w}_j. \tag{8.9}$$

In matrix form, we can rewrite Eq. (8.9) as

$$\mathbf{Z}^{\mathrm{a}} = \mathbf{Z}^{\mathrm{f}} + \mathbf{A}\mathbf{W}, \tag{8.10}$$

where column j of $\mathbf{W} \in \Re^{N \times N}$ is just $\mathbf{w}_j$ from Eq. (8.9).

Following Hunt et al. (2007) we write the cost function (7.1) in terms of $\mathbf{w}_j$ as

---

Cost function in ensemble subspace

$$\mathcal{J}(\mathbf{w}_j) = \frac{1}{2}\mathbf{w}_j^{\mathrm{T}}\mathbf{w}_j + \frac{1}{2}\left(\mathbf{g}(\mathbf{z}_j^{\mathrm{f}} + \mathbf{A}\mathbf{w}_j) - \mathbf{d}_j\right)^{\mathrm{T}}\overline{\mathbf{C}}_{dd}^{-1}\left(\mathbf{g}(\mathbf{z}_j^{\mathrm{f}} + \mathbf{A}\mathbf{w}_j) - \mathbf{d}_j\right), \tag{8.11}$$

---

where we have used that

$$\begin{aligned}
\mathbf{w}_j^{\mathrm{T}}\mathbf{A}^{\mathrm{T}}\mathbf{C}_{zz}^{-1}\mathbf{A}\mathbf{w}_j &\approx \mathbf{w}_j^{\mathrm{T}}\mathbf{A}^{\mathrm{T}}\overline{\mathbf{C}}_z^{-1}\mathbf{A}\mathbf{w}_j \\
&= \mathbf{w}_j^{\mathrm{T}}\mathbf{A}^{\mathrm{T}}\left(\mathbf{A}\mathbf{A}^{\mathrm{T}}\right)^{\dagger}\mathbf{A}\mathbf{w}_j \\
&= \mathbf{w}_j^{\mathrm{T}}\left(\mathbf{A}^{\dagger}\mathbf{A}\right)\mathbf{w}_j \\
&= \mathbf{w}_j^{\mathrm{T}}\left(\mathbf{A}^{\dagger}\mathbf{A}\right)\left(\mathbf{A}^{\dagger}\mathbf{A}\right)\mathbf{w}_j \\
&= \widetilde{\mathbf{w}}_j^{\mathrm{T}}\widetilde{\mathbf{w}}_j.
\end{aligned} \tag{8.12}$$

in which we defined $\widetilde{\mathbf{w}}_j = \left(\mathbf{A}^{\dagger}\mathbf{A}\right)\mathbf{w}_j$. The superscript $\dagger$ denotes the pseudo inverse. The expression $\mathbf{A}^{\dagger}\mathbf{A}$ is the orthogonal projection onto the range of $\mathbf{A}^{\mathrm{T}}$. Also, we have the projection property $\left(\mathbf{A}^{\dagger}\mathbf{A}\right)\left(\mathbf{A}^{\dagger}\mathbf{A}\right) = \mathbf{A}^{\dagger}\mathbf{A}$. Thus, $\widetilde{\mathbf{w}}_j$ is just the projection of $\mathbf{w}$ onto ensemble perturbation space, and from

$$\mathbf{A}\mathbf{w}_j = \mathbf{A}\left(\mathbf{A}^{\dagger}\mathbf{A}\right)\mathbf{w}_j = \mathbf{A}\widetilde{\mathbf{w}}_j, \tag{8.13}$$

we see that it does not matter whether we solve for $\mathbf{w}_j$ or $\widetilde{\mathbf{w}}_j$.

Note that we have used an ensemble representation for the measurement-error-covariance matrix. We could have retained the complete $\mathbf{C}_{dd}$, but in many cases, this matrix is too large for practical computations, and it is therefore commonly approximated by a diagonal matrix. Thus, one typically neglects all measurement error correlations, which can have dire consequences. Evensen (2021) proposed using the ensemble representation in Eq. (8.6), but with an increased ensemble size to mitigate additional sampling errors. We will discuss this issue further in Chap. 13.

Minimizing the cost functions in Eq. (8.11) implies solving for the minima of the original cost functions in Eq. (7.1), but restricted to the ensemble subspace and with $\overline{\mathbf{C}}_{\overline{z}}$ in place of $\mathbf{C}_{zz}$, as explained by Bocquet et al. (2015). The ensemble of cost functions in Eq. (8.11) does not refer to the high-dimensional state-covariance matrix, $\mathbf{C}_{zz}$. In the original formulation, we searched for the solution in the state space. We now have a simpler problem where we search for the ensemble sub-space solution. Thus, we solve for the N vectors $\mathbf{w}_j \in \mathfrak{R}^N$, one for each realization.

8.4 Ensemble Subspace RML

We will formulate a Gauss–Newton method for minimizing the cost function in Eq. (8.11) and the following algorithm comes from Evensen et al. (2019). The Jacobian (gradient) of the cost function $\nabla_{\mathbf{w}}\mathcal{J}(\mathbf{w}_j) \in \mathfrak{R}^{N \times 1}$ is

$$\nabla_{\mathbf{w}}\mathcal{J}(\mathbf{w}_j) = \mathbf{w}_j + (\mathbf{G}_j\mathbf{A})^{\mathrm{T}}\overline{\mathbf{C}}_{dd}^{-1}\big(\mathbf{g}(\mathbf{z}_j^{\mathrm{f}} + \mathbf{A}\mathbf{w}_j) - \mathbf{d}_j\big), \qquad (8.14)$$

and an approximate Hessian (gradient of the Jacobian) $\nabla_{\mathbf{w}}\nabla_{\mathbf{w}}\mathcal{J}(\mathbf{W}) \in \mathfrak{R}^{N \times N}$ becomes

$$\nabla_{\mathbf{w}}\nabla_{\mathbf{w}}\mathcal{J}(\mathbf{w}_j) \approx \mathbf{I} + (\mathbf{G}_j\mathbf{A})^{\mathrm{T}}\overline{\mathbf{C}}_{dd}^{-1}(\mathbf{G}_j\mathbf{A}). \qquad (8.15)$$

We have defined the tangent-linear model

$$\mathbf{G}_j = \big(\nabla_{\mathbf{w}}\mathbf{g}|_{\mathbf{z}_j^{\mathrm{f}} + \mathbf{A}\mathbf{w}_j}\big)^{\mathrm{T}} \in \mathfrak{R}^{m \times n}, \qquad (8.16)$$

and in the Hessian we have neglected the second-order derivatives.

The iterative Gauss–Newton scheme for minimizing the cost function in Eq. (8.11), analogous to (7.5), is then

$$\begin{aligned}
\mathbf{w}_j^{i+1} = \mathbf{w}_j^i - \gamma &\left(\mathbf{I} + (\mathbf{G}_j^i\mathbf{A})^{\mathrm{T}}\overline{\mathbf{C}}_{dd}^{-1}(\mathbf{G}_j^i\mathbf{A})\right)^{-1} \\
&\times \left(\mathbf{w}_j^i + (\mathbf{G}_j^i\mathbf{A})^{\mathrm{T}}\overline{\mathbf{C}}_{dd}^{-1}\big(\mathbf{g}(\mathbf{z}_j^{\mathrm{f}} + \mathbf{A}\mathbf{w}_j^i) - \mathbf{d}_j\big)\right),
\end{aligned} \qquad (8.17)$$

where we introduce $\gamma \in (0, 1]$ as a step-length parameter, and we have the tangent-linear operator evaluated for realization j at the current iteration i as

$$\mathbf{G}_j^i = \big(\nabla_{\mathbf{w}}\mathbf{g}|_{\mathbf{z}_j^{\mathrm{f}} + \mathbf{A}\mathbf{w}_j^i}\big)^{\mathrm{T}}. \qquad (8.18)$$

Now using the corollaries from Eqs. (6.9) and (6.10), we can write the Gauss–Newton iteration in Eq. (8.17) as

$$\begin{aligned}
\mathbf{w}_j^{i+1} = \mathbf{w}_j^i - \gamma &\left\{\mathbf{w}_j^i - (\mathbf{G}_j^i\mathbf{A})^{\mathrm{T}}\left((\mathbf{G}_j^i\mathbf{A})(\mathbf{G}_j^i\mathbf{A})^{\mathrm{T}} + \overline{\mathbf{C}}_{dd}\right)^{-1}\right. \\
&\times \left.\left((\mathbf{G}_j^i\mathbf{A})\mathbf{w}_j^i + \mathbf{d}_j - \mathbf{g}(\mathbf{z}_j^{\mathrm{f}} + \mathbf{A}\mathbf{w}_j^i)\right)\right\}.
\end{aligned} \qquad (8.19)$$

By introducing the ensemble representation for the covariances in the linear regression we obtain

$$\mathbf{G}_j^i \approx \overline{\mathbf{G}}^i \triangleq \overline{\mathbf{C}}_{yz}^i \overline{\mathbf{C}}_{zz}^{i\,\dagger} = \mathbf{Y}^i \mathbf{A}^{i\,\dagger}, \tag{8.20}$$

where $\mathbf{Y}^i$ is defined from Eq. (8.21) and evaluated at iteration i, i.e.,

$$\mathbf{Y}^i = \mathbf{g}(\mathbf{Z}^i)\mathbf{\Pi}. \tag{8.21}$$

The tricky term in Eq. (8.19), which correponds to the one mentioned in relation to Eq. (7.9), is the product $\mathbf{G}_j^i \mathbf{A}$. Evensen et al. (2019) showed that we can write

$$\mathbf{S}^i = \overline{\mathbf{G}}^i \mathbf{A} = \mathbf{Y}^i \mathbf{A}^{i\,\dagger} \mathbf{A} \tag{8.22}$$

$$= \mathbf{Y}^i \mathbf{A}^{i\,\dagger} \mathbf{A}^i \mathbf{\Omega}^{i\,-1} \tag{8.23}$$

$$= \mathbf{Y}^i \mathbf{\Omega}^{i\,-1} \quad \text{if } n \geq N-1 \text{ or if } \mathbf{g} \text{ is linear.} \tag{8.24}$$

In this expression, we have defined the quadratic matrix

$$\mathbf{\Omega}^i = \mathbf{I} + \mathbf{W}^i \mathbf{\Pi}, \tag{8.25}$$

that relates the ensemble anomalies at iteration i to the initial anomalies $\mathbf{A} = \mathbf{A}^i \mathbf{\Omega}^{i-1}$.

Note that we cannot use Eq. (8.24) when $n < N-1$, i.e., when the state dimension is less that the ensemble size minus one. We, then, need to retain the projection $\mathbf{A}^{i\,\dagger} \mathbf{A}^i$ and use Eq. (8.23) rather than Eq. (8.24). Evensen et al. (2019) derived the proofs of this result and we refer to this paper for the details. This result is also complementary to Eq. (7.11) which was derived by Evensen (2019).

We can now write the iteration of Eq. (8.19) in matrix form as

$$\mathbf{W}^{i+1} = \mathbf{W}^i - \gamma \left(\mathbf{W}^i - \mathbf{S}^{i\,\mathrm{T}} \left(\mathbf{S}^i \mathbf{S}^{i\,\mathrm{T}} + \overline{\mathbf{C}}_{dd} \right)^{-1} \widetilde{\mathbf{D}}^i \right), \tag{8.26}$$

where we have defined the "innovation" term

$$\widetilde{\mathbf{D}}^i = \mathbf{S}^i \mathbf{W}^i + \mathbf{D} - \mathbf{g}(\mathbf{Z}^i). \tag{8.27}$$

The update for iteration i is

$$\mathbf{Z}^i = \mathbf{Z} + \mathbf{A}\mathbf{W}^i$$

$$= \mathbf{Z}\left(\mathbf{I} + \mathbf{\Pi}\mathbf{W}^i / \sqrt{N-1} \right) \tag{8.28}$$

$$= \mathbf{Z}\left(\mathbf{I} + \mathbf{W}^i / \sqrt{N-1} \right),$$

where we have used $\mathbf{W}^i = \mathbf{\Pi}\mathbf{W}^i$, which we get from Eq. (8.26) using $\mathbf{S}^{\mathrm{T}} = \mathbf{\Pi}\mathbf{S}^{\mathrm{T}}$. Thus, we can compute the final update to a cost of nN^2 operations. The updated ensemble is a linear combination of the prior ensemble members, and the prior ensemble space contains the updated ensemble of solutions. Algorithm 5 details the implementation of the ensemble subspace RML algorithm. The algorithm takes as inputs the prior ensemble and the perturbed measurements, and runs an ensemble of model simulations to evaluate $\mathbf{g}(\mathbf{Z}^i)$. Thus, the algorithm is generic and we can use it for any model or problem configuration. In Sect. 8.9 we discuss a practical and efficient implementation for inverting the expression $\left(\mathbf{S}^i \mathbf{S}^{i\,\mathrm{T}} + \overline{\mathbf{C}}_{dd} \right)^{-1}$ where we replace the full measurement error covariance matrix with the ensemble representation, $\mathbf{C}_{dd} \approx \overline{\mathbf{C}}_{dd} = \mathbf{E}\mathbf{E}^{\mathrm{T}}$. We have used this subspace EnRML method in the petroleum example in Chap. 21.

Algorithm 5 Subspace EnRML algorithm for one assimilation window

---

1: Input: $\mathbf{Z} \in \mathfrak{R}^{n \times N}$ ▷ Prior state-vector ensemble
2: Input: $\mathbf{D} \in \mathfrak{R}^{m \times N}$ ▷ Perturbed measurements
3: $\mathbf{W}^{(0)} = 0$ ▷ $\mathbf{W} \in \mathfrak{R}^{N \times N}$
4: $\mathbf{\Pi} = \left(\mathbf{I} - \frac{1}{N}\mathbf{1}\mathbf{1}^{\mathrm{T}}\right)/\sqrt{N-1}$ ▷ $\mathbf{\Pi} \in \mathfrak{R}^{N \times N}$
5: $\mathbf{E} = \mathbf{D}\mathbf{\Pi}$ ▷ $\mathbf{E} \in \mathfrak{R}^{m \times N}$
6: $i = 0$
7: **repeat**
8:    $\mathbf{Y}^i = \mathbf{g}(\mathbf{Z}^i)\mathbf{\Pi}$ ▷ $\mathbf{Y} \in \mathfrak{R}^{m \times N}$
9:    **if** $n < N - 1$ **then**
10:       $\mathbf{Y}^i = \mathbf{Y}^i \mathbf{A}^{i\dagger} \mathbf{A}^i$
11:    **end if**
12:    $\mathbf{\Omega}^i = \mathbf{I} + \mathbf{W}^i \mathbf{\Pi}$ ▷ $\mathbf{\Omega} \in \mathfrak{R}^{N \times N}$
13:    $\mathbf{S}^i = \mathbf{Y}^i \mathbf{\Omega}^{i^{-1}}$ ▷ $\mathbf{S} \in \mathfrak{R}^{m \times N}$
14:    $\widetilde{\mathbf{D}}^i = \mathbf{S}^i \mathbf{W}^i + \mathbf{D} - \mathbf{g}(\mathbf{Z}^i)$ ▷ $\widetilde{\mathbf{D}} \in \mathfrak{R}^{m \times N}$
15:    $\mathbf{W}^{i+1} = \mathbf{W}^i - \gamma\left(\mathbf{W}^i - \mathbf{S}^{i^{\mathrm{T}}}\left(\mathbf{S}^i \mathbf{S}^{i^{\mathrm{T}}} + \mathbf{E}\mathbf{E}^{\mathrm{T}}\right)^{-1}\widetilde{\mathbf{D}}^i\right)$
16:    $\mathbf{T}^i = \left(\mathbf{I} + \mathbf{W}^{i+1}/\sqrt{N-1}\right)$ ▷ $\mathbf{T} \in \mathfrak{R}^{N \times N}$
17:    $\mathbf{Z}^{i+1} = \mathbf{Z}\mathbf{T}^i$
18:    $i = i + 1$
19: **until** convergence

8.5 Ensemble Kalman Filter (EnKF) Update

We can derive the EnKF update as a minimizer of the ensemble of cost functions in Eq. (7.1). For this, we compute the solution from Eq. (7.12), but using the ensemble of realizations in Eq. (8.1) to represent the error covariance matrix. Thus, we can write

$$\mathbf{z}_j^{\mathrm{a}} = \mathbf{z}_j^{\mathrm{f}} + \overline{\mathbf{C}}_{zy}\left(\overline{\mathbf{C}}_{yz}\overline{\mathbf{C}}_{zz}^{\dagger}\overline{\mathbf{C}}_{zy} + \overline{\mathbf{C}}_{dd}\right)^{-1}\left(\mathbf{d}_j - \mathbf{g}(\mathbf{z}_j^{\mathrm{f}})\right). \tag{8.29}$$

Equation (8.29) represents the EnKF update equation of Evensen (1994) with the perturbed observations proposed by Burgers et al. (1998).

It is straightforward to show that

$$\begin{aligned}
\overline{\mathbf{C}}_{yz}\overline{\mathbf{C}}_{zz}^{\dagger}\overline{\mathbf{C}}_{zy} &= \mathbf{Y}\mathbf{A}^{\mathrm{T}}\left(\mathbf{A}\mathbf{A}^{\mathrm{T}}\right)^{\dagger}\mathbf{A}\mathbf{Y}^{\mathrm{T}} \\
&= \mathbf{Y}\left(\mathbf{A}^{\dagger}\mathbf{A}\right)\left(\mathbf{Y}\left(\mathbf{A}^{\dagger}\mathbf{A}\right)\right)^{\mathrm{T}} \\
&= \overline{\mathbf{C}}_{yy} \quad \text{for } n \geq N - 1,
\end{aligned} \tag{8.30}$$

by using the following result from Sakov et al. (2012),

$$\mathbf{A}^{\dagger}\mathbf{A} = \mathbf{I}_N - \frac{1}{N}\mathbf{1}\mathbf{1}^{\mathrm{T}} \quad \text{for } n > N - 1. \tag{8.31}$$

However, only in the low-rank case, when $n > N - 1$, is $\mathbf{A}^{\dagger}\mathbf{A}$ a projection that removes the ensemble mean as defined in Eq. (8.12). But, since in Eq. (8.30), the

---

Algorithm 6 Subspace EnKF update

1: subroutine EnKF_update$(\mathbf{Z}, \mathbf{D}, \mathbf{\Upsilon})$
2: Input: $\mathbf{Z} \in \mathfrak{R}^{n \times N}$ ▷ Prior state-vector ensemble
3: Input: $\mathbf{D} \in \mathfrak{R}^{m \times N}$ ▷ Perturbed measurements
4: Input: $\mathbf{\Upsilon} \in \mathfrak{R}^{m \times N}$ ▷ Predicted measurements
5: $\mathbf{\Pi} = \left(\mathbf{I} - \frac{1}{N}\mathbf{1}\mathbf{1}^T\right)/\sqrt{N-1}$ ▷ $\mathbf{\Pi} \in \mathfrak{R}^{N \times N}$
6: $\mathbf{E} = \mathbf{D}\mathbf{\Pi}$ ▷ $\mathbf{E} \in \mathfrak{R}^{m \times N}$
7: $\mathbf{Y} = \mathbf{\Upsilon}\mathbf{\Pi}$ ▷ $\mathbf{Y} \in \mathfrak{R}^{m \times N}$
8: **if** $n < N - 1$ **then**
9:     $\mathbf{Y} = \mathbf{Y}\mathbf{A}^\dagger\mathbf{A}$
10: **end if**
11: $\mathbf{S} = \mathbf{Y}$ ▷ $\mathbf{S} \in \mathfrak{R}^{m \times N}$
12: $\widetilde{\mathbf{D}} = \mathbf{D} - \mathbf{\Upsilon}$ ▷ $\widetilde{\mathbf{D}} \in \mathfrak{R}^{m \times N}$
13: $\mathbf{W} = \mathbf{S}^T\left(\mathbf{S}\mathbf{S}^T + \mathbf{E}\mathbf{E}^T\right)^{-1}\widetilde{\mathbf{D}}$ ▷ $\mathbf{W} \in \mathfrak{R}^{N \times N}$
14: $\mathbf{T} = \left(\mathbf{I} + \mathbf{W}/\sqrt{N-1}\right)$ ▷ $\mathbf{T} \in \mathfrak{R}^{N \times N}$
15: $\mathbf{Z} \leftarrow \mathbf{Z}\mathbf{T}$ ▷ Update returned in $\mathbf{Z}$

---

mean of $\mathbf{Y}$ is already zero by the definition in Eq. (8.8), the additional multiplication with $\mathbf{A}^\dagger\mathbf{A}$ has no effect.

Evensen (2003) reformulated the EnKF update Eq. (8.29) in terms of the ensemble, as

$$\begin{aligned}\mathbf{Z}^a &= \mathbf{Z}^f + \mathbf{A}\mathbf{Y}^T\left(\mathbf{Y}\mathbf{Y}^T + \mathbf{E}\mathbf{E}^T\right)^{-1}\left(\mathbf{D} - \mathbf{g}(\mathbf{Z}^f)\right) \\ &= \mathbf{Z}^f + \mathbf{A}\mathbf{W},\end{aligned} \tag{8.32}$$

with

$$\mathbf{W} = \mathbf{Y}^T\left(\mathbf{Y}\mathbf{Y}^T + \mathbf{E}\mathbf{E}^T\right)^{-1}\left(\mathbf{D} - \mathbf{g}(\mathbf{Z}^f)\right), \tag{8.33}$$

but he did not realize the limitation in Eq. (8.30). Thus, for the Eq. (8.32) to be generally valid, we need to redefine $\mathbf{Y}$ as follows

$$\mathbf{Y} = \begin{cases} \mathbf{Y} & \text{for } n \geq N - 1 \\ \mathbf{Y}\mathbf{A}^\dagger\mathbf{A} & \text{for } n < N - 1. \end{cases} \tag{8.34}$$

Interestingly, it is possible to compute the EnKF solution from the first ensemble-subspace RML iteration. The prior value of $\mathbf{W}$ is $\mathbf{W}^{(0)} = 0$, and if we set the step-length $\gamma = 1.0$, then the first Gauss–Newton iteration of (8.26) becomes just the EnKF update equation. Hence, the EnKF solution is also confined to the prior ensemble subspace. Moreover, if we implement Algorithm 5, we can use it to compute both the EnRML and the EnKF solutions. We have presented a simplified EnKF computation in Algorithm 6.

Let us define $\mathbf{X}$ where each column stores an ensemble realization for all of the K time steps of the window. This notation allows us to write $\mathbf{X}_k$ to represent the rows in $\mathbf{X}$ holding the solution corresponding to time step k. But more importantly, we can

---

Algorithm 7 Recursive EnKF updates using various filter and smoother configurations

---

1: Input: $\mathbf{Z} \in \mathfrak{R}^{n \times N}$ ▷ Initial model state-vector ensemble
2: Input: $\mathbf{D}_l \in \mathfrak{R}^{m \times N}$ ▷ Perturbed measurements for each assimilation window
3: Input: analysis   ▷ Analysis method
4: **for** $l = 1, \ldots$ **do**   ▷ Loop over assimilation windows
5:     $\mathbf{X}_0 = \mathbf{Z}$
6:     **for** $k = 1, K$ **do**   ▷ Ensemble integration
7:         $\mathbf{X}_k = \mathbf{m}(\mathbf{X}_{k-1})$
8:     **end for**
9:     $\mathbf{X} = [\mathbf{X}_0, \ldots, \mathbf{X}_K]$
10:    $\Upsilon = \mathbf{h}(\mathbf{X})$ ▷ Predicted measurments
11:    **select case (analysis)**
12:        **case (EnKF)**   ▷ Recursive EnKF updates at end of assimilation window
13:            **call EnKF_update**$(\mathbf{X}_K, \mathbf{D}_l, \Upsilon)$

14:        **case (EnKF0)**   ▷ Recursive EnKF updates at start of assimilation window
15:            **call EnKF_update**$(\mathbf{X}_0, \mathbf{D}_l, \Upsilon)$
16:            **for** $k = 1, K$ **do**   ▷ Rerun ensemble integration
17:                $\mathbf{X}_k = \mathbf{m}(\mathbf{X}_{k-1})$
18:            **end for**

19:        **case (ES)**   ▷ Recursive updates over the whole window
20:            **call EnKF_update**$(\mathbf{X}, \mathbf{D}_l, \Upsilon)$

21:        **case (EnKS)**   ▷ Recursive smoother updates over all previous windows
22:            $\mathcal{X}_l = [\mathcal{X}_{l-1}, \mathbf{X}]$
23:            **call EnKF_update**$(\mathcal{X}_l, \mathbf{D}_l, \Upsilon)$
24:            $\mathbf{X} = \mathcal{X}_l(l)$

25:        **case (EnKS_lagged)**   ▷ Recursive smoother updates over λ previous windows
26:            $\mathcal{X}_l = [\mathcal{X}_{l-1}(l - \lambda : l - 1), \mathbf{X}]$
27:            **call EnKF_update**$(\mathcal{X}_l, \mathbf{D}_l, \Upsilon)$
28:            $\mathbf{X} = \mathcal{X}_l(l)$
29:    **end select case**
30:    $\mathbf{Z} = \mathbf{X}_K$ ▷ Define state vector for next assimilation window
31: **end for**

---

have measurements distributed over the assimilation window, and we can compute the predicted measurements just by measuring $\mathbf{X}$, i.e., $\Upsilon = \mathbf{h}(\mathbf{X})$.

Note that by using this formulation of the EnKF, we have complete flexibility in defining what the state vector is (see Algorithm 7). It allows us to update the model state at the end of the assimilation window, as is common in sequential data assimilation using EnKF (see also the filter solution in Fig. 2.2). And it also allows us to use measurements distributed over the assimilation interval in the update calculation.

But, this formulation also allows us to compute the solution at the initial time of the assimilation window and then integrate the posterior ensemble over the assimilation window to obtain the prior for the next window as in Fig. 2.4. Or we can update the model state $\mathbf{X}$ over the whole assimilation window, which corresponds to the ensemble smoother (ES) solution in Fig. 2.1 as introduced in Van Leeuwen and Evensen (1996) and first applied to a real oceanographic applications in Van Leeuwen (1999, 2001). Finally, the formulation is flexible enough to augment the updates from the previous windows to the state vector and recursively update the model state in the current and some previous windows. This last approach corresponds to the (lagged) ensemble Kalman smoother (EnKS), as introduced in Evensen and Van Leeuwen (2000). We illustrate all these alternatives in Algorithm 7, where the main difference between them is the definition of the state vector we are updating.

A final remark is that the original Kalman filter usually writes the update equations using the Kalman gain matrix $\mathbf{K}$. However, it only makes sense to compute $\mathbf{K}$ when one has a full-rank error covariance matrix $\mathbf{C}_{zz}$. In the ensemble methods, the use of low-rank representation $\overline{\mathbf{C}}_{zz}$ implies that $\mathbf{K}$ is also of low rank as we compute an ensemble representation, $\overline{\mathbf{K}}$, from an outer product of low-rank ensemble matrices. In other words, one should never compute the Kalman gain matrix using ensemble methods unless maybe when the number of measurements is less than the ensemble size.

8.6 Ensemble DA with Multiple Updating (ESMDA)

For some non-linear problems, the Gauss–Newton method may not converge if the normalizing Hessian is of low rank. We can then use an alternative formulation named ensemble smoother with multiple data assimilation (ESMDA) proposed by Emerick and Reynolds (2013). ESMDA approximately samples from $f(\mathbf{z}|\mathbf{d})$ by gradually introducing measurements using the so-called tapering of the likelihood function (Neal, 1996).

When requiring that

$$\sum_{i=1}^{N_{\mathrm{mda}}} \frac{1}{\alpha^i} = 1, \tag{8.35}$$

we can write the following

$$
\begin{aligned}
f(\mathbf{z}|\mathbf{d}) &\propto f(\mathbf{d}|\mathbf{g}(\mathbf{z}))\, f(\mathbf{z}) \\
&= f(\mathbf{d}|\mathbf{g}(\mathbf{z}))^{\left(\sum_{i=1}^{N_{\mathrm{mda}}} \frac{1}{\alpha^i}\right)} f(\mathbf{z}) \\
&= f(\mathbf{d}|\mathbf{g}(\mathbf{z}))^{\frac{1}{\alpha^{N_{\mathrm{mda}}}}} \cdots f(\mathbf{d}\,|\,\mathbf{g}(\mathbf{z}))^{\frac{1}{\alpha^2}} f(\mathbf{d}\,|\,\mathbf{g}(\mathbf{z}))^{\frac{1}{\alpha^1}} f(\mathbf{z}).
\end{aligned} \tag{8.36}
$$

We can then compute N_{mda} recursive EnKF steps that gradually introduce the observations using inflated observation errors. This gradual introduction of the update reduces the impact of the linearization in the ES scheme, see Approx. 5. The method

---

Algorithm 8 ESMDA iterations for an assimilation window

1: Input: $\mathbf{Z} \in \mathfrak{R}^{n \times N}$ ▷ Initial model state-vector ensemble
2: Input: $\mathbf{d} \in \mathfrak{R}^{m}$ ▷ Measurements within the window
3: Input: $\mathbf{C}_{dd} \in \mathfrak{R}^{m \times m}$ ▷ Measurement error covariance
4: **for** $l = 1, \ldots$ **do**   ▷ Loop over assimilation windows
5:     **for** $i = 1, N_{\mathrm{mda}}$ **do**   ▷ Loop over assimilation windows
6:       $\mathbf{X}_0 = \mathbf{Z}$
7:       **for** $k = 1, K$ **do**   ▷ Ensemble integration
8:         $\mathbf{X}_k = \mathbf{m}(\mathbf{X}_{k-1})$
9:       **end for**
10:      $\mathbf{X} = (\mathbf{X}_0, \ldots, \mathbf{X}_K)$
11:      $\mathbf{\Upsilon} = \mathbf{h}(\mathbf{X})$ ▷ Predicted measurments
12:      $\mathbf{E} = \mathcal{N}(0, \mathbf{C}_{dd})$ ▷ Resample new meaurement perturbations
13:      $\mathbf{D} = \mathbf{d}\mathbf{1}^{\mathrm{T}} + \sqrt{\alpha}\mathbf{E}$ ▷ Create new scaled perturbed measurements
14:      **call EnKF_update** $\left(\mathbf{X}_0, \mathbf{D}_l, \mathbf{\Upsilon}\right)$ ▷ Compute update at start of window
15:    **end for**
16:    **for** $k = 1, K$ **do**   ▷ Rerun ensemble integration
17:      $\mathbf{X}_k = \mathbf{m}(\mathbf{X}_{k-1})$
18:    **end for**
19:    $\mathbf{Z} = \mathbf{X}_N$ ▷ Define state vector for next window
20: **end for**

---

converges precisely to the ES solution in the linear case when the ensemble size goes to infinity.

In ESMDA, we can use Algorithm 6 to compute the solution. We follow each step of the algorithm as we do for the EnKF solution, but we repeat the procedure N_{mda} times. For each recursive call to the algorithm, we will resample the perturbed measurements from $\mathcal{N}(\mathbf{d}, \alpha^i \overline{\mathbf{C}}_{dd})$. Thus, the effective measurement error variance in each step is increased with a factor α^i.

ESMDA has gained popularity due to its ease of implementation and its successful use in various applications. Although it is unclear what the method converges to in the nonlinear case, it appears to provide an acceptable solution in many cases. Note that ESMDA with one step corresponds to the EnKF estimate for the start of the assimilation window.

When we consider the convergence of ESMDA, we mean the number of steps needed before a further decrease in step length does not change the final solution. The required number of steps depends on the nonlinearity of the model. In the COVID example in Chap. 22, we found that 16–32 steps were necessary. We note that in ESMDA, as the number of steps increases, the measurement perturbations also increase. Thus, one can imagine cases where the perturbed measurements take unphysical values causing the algorithm to break down. Emerick (2018) resolved this particular issue by using a square-root formulation for the update calculation.

To summarize, both EnRML and ESMDA solve the recursive smoother problem over a sequence of assimilation windows. It is unclear which method will work the best for a particular situation, and both have their advantages and disadvantages. EnKF-type approaches are more efficient to compute as they linearize the measurement prediction and avoid iterations. However, as we will see in the following section, we can also use the 4DVar methods in an ensemble setting and compute the ensemble update without applying the linear regression of Approx. 7.

8.7   Ensemble 4DVar with Consistent Error Statistics

In Chaps. 4 and 5, we learned that the 4DVar method solves for the maximum a posteriori solution if it converges to the global minimum of the cost function in Eq. (3.9). Thus, we can use 4DVar to minimize the ensemble of cost functions in Eq. (7.1). These cost functions are all independent of each other. Suppose the operator $\mathbf{g}(\mathbf{z})$ is weakly nonlinear. In that case, we can minimize each cost function independently using 4DVar to obtain an ensemble of solutions representing the minima defined by the cost functions in Eq. (7.1). This approach lets us sample approximately the marginal probability in Bayes' theorem in Eq. (2.43). This approach of sampling realizations from the posterior pdf, referred to as the *Ensemble of Data Assimilations' (En4DVar)* approach (Isaksen et al., 2010), is an example of RML sampling and is the method currently used at ECMWF for operational data assimilation.

En4DVar samples an ensemble of realizations from the posterior pdf, so it can also represent the posterior error covariance matrix. Using SC-4DVar, we sample the ensemble of initial conditions for the time window and obtain the ensemble solution over the time window by a final ensemble integration. At the end of the time window, the ensemble prediction provides the initial conditions for the next assimilation window, and the ensemble spread represents its updated background error-covariance matrix.

The WC-4DVar solution of En4DVar gives an updated ensemble for the whole assimilation window. While SC-En4DVar initializes a prediction from the estimate at the beginning of the assimilation window, when using WC-En4DVar, we should initialize the forecast from the ensemble estimate at the end of the assimilation window. Like SC-En4DVar, the posterior ensemble of WC-En4DVar solutions also represents the posterior error covariances and the background error covariance for the next assimilation window.

Hence, we can use both SC-En4DVar and WC-En4DVar to use the ensemble statistics in recursive model updating as we do when using EnKF. The advantage of En4DVar is that we minimize precisely the cost functions in Eq. (7.1) without using the ensemble-averaged model sensitivity as in EnKF. In practice, the ensemble size is small for high-dimensional 4DVar applications, typically much less than 100 members. In this case, the ensemble covariance matrix is a poor estimate of the prior covariance matrix in a 4Dvar, which degrades the assimilation results. A partial solution to this problem is to use the ensemble to update only the variances and a few length scales in a climatological prior covariance matrix.

Note that the issue of correctly representing the error-covariance matrix also exists in the EnKF. But, since the computational cost of EnKF is much less than for an ensemble 4DVar, we can partly resolve this problem by using a larger ensemble size with EnKF. Additionally, we often use the localization and inflation schemes, as discussed in Chap. 10.

8.8 Square-Root EnKF

The so-called square-root filters belong to a popular class of ensemble Kalman filters. These are ensemble filters that do not attempt to sample the Bayesian posterior pdf. Instead, the square root methods assume a Gaussian posterior distribution with a covariance matrix equal to the Kalman-filter analysis covariance matrix in Eq. (6.33). The square-root filters' popularity comes from their avoidance of using perturbed observations, reducing sampling errors. We can use different routes of deriving the square-root update equation, but one most commonly starts from a factorization of Eq. (6.33) using ensemble covariances.

Of course, we usually do not know the analysis error covariance, and if we knew it, we would not be able to factorize it for many real-sized applications. On the other hand, when using ensemble methods, we can replace the covariances in Eq. (6.33) with their ensemble representations, i.e.,

$$\mathbf{A}^{a}\mathbf{A}^{a\mathrm{T}} = \mathbf{A}\mathbf{A}^{\mathrm{T}} - \mathbf{A}\mathbf{Y}\left(\mathbf{Y}\mathbf{Y}^{\mathrm{T}} + \mathbf{C}_{dd}\right)^{-1}\mathbf{Y}\mathbf{A}^{\mathrm{T}} \tag{8.37}$$

$$= \mathbf{A}\left(\mathbf{I} - \mathbf{Y}\mathbf{C}^{-1}\mathbf{Y}\right)\mathbf{A}^{\mathrm{T}} \tag{8.38}$$

$$= \mathbf{A}\left(\mathbf{Q}\mathbf{\Lambda}\mathbf{Q}^{\mathrm{T}}\right)\mathbf{A}^{\mathrm{T}} \tag{8.39}$$

$$= \left(\mathbf{A}\mathbf{Q}\mathbf{\Lambda}^{\frac{1}{2}}\right)\left(\mathbf{A}\mathbf{Q}\mathbf{\Lambda}^{\frac{1}{2}}\right)^{\mathrm{T}} \tag{8.40}$$

$$= \left(\mathbf{A}\mathbf{Q}\mathbf{\Lambda}^{\frac{1}{2}}\mathbf{Q}^{\mathrm{T}}\right)\left(\mathbf{A}\mathbf{Q}\mathbf{\Lambda}^{\frac{1}{2}}\mathbf{Q}^{\mathrm{T}}\right)^{\mathrm{T}} \tag{8.41}$$

$$= \left(\mathbf{A}\mathbf{Q}\mathbf{\Lambda}^{\frac{1}{2}}\mathbf{Q}^{\mathrm{T}}\mathbf{\Theta}\right)\left(\mathbf{A}\mathbf{Q}\mathbf{\Lambda}^{\frac{1}{2}}\mathbf{Q}^{\mathrm{T}}\mathbf{\Theta}\right)^{\mathrm{T}}. \tag{8.42}$$

In Eq. (8.37), we have used the definitions in Eqs. (8.4), (8.21), and (8.34) to rewrite the Kalman filter error covariance update of Eq. (6.33) using the ensemble matrices. In Eq. (8.38), we define the matrix $\mathbf{C} = \mathbf{Y}\mathbf{Y}^{\mathrm{T}} + \mathbf{C}_{dd}$. After that, in Eq. (8.39), we use the eigendecomposition

$$\mathbf{Q}\mathbf{\Lambda}\mathbf{Q}^{\mathrm{T}} = \left(\mathbf{I} - \mathbf{Y}\mathbf{C}^{-1}\mathbf{Y}\right), \tag{8.43}$$

where we factorize a matrix of the dimension the number of ensemble members N. To make the method even more efficient, in Sect. 8.9, we will present an algorithm that computes the inversion of $\mathbf{C}$ in the ensemble subspace of dimension N.

One alternative for updating the forecast ensemble anomalies is to use (8.40) and write

$$\mathbf{A}^{\mathrm{a}} = \mathbf{A}\mathbf{Q}\boldsymbol{\Lambda}^{\frac{1}{2}}, \tag{8.44}$$

which we usually refer to as the one-sided square-root update. Evensen (2009b) and Leeuwenburgh (2005) showed that this asymmetrical scheme leads to a solution that does not conserve the mean, and it creates outliers that hold most of the variance. However, by using the symmetric square root in Eq. (8.40), we obtain an update-equation

$$\mathbf{A}^{\mathrm{a}} = \mathbf{A}\mathbf{Q}\boldsymbol{\Lambda}^{\frac{1}{2}}\mathbf{Q}^{\mathrm{T}}, \tag{8.45}$$

that ensures zero mean for the anomalies. The updated ensemble becomes a "symmetric" and "scaled" contraction along the different eigenvectors in $\mathbf{Q}$. If we desire a more randomized update, we can add the orthogonal random matrix $\boldsymbol{\Theta}$, which randomizes the anomaly updates among the different directions in the eigenvector space.

Several publications show the superiority of the square-root schemes for low dimensional models and with small ensemble sizes of $O(10)$. In these examples, the square-root solution becomes nearly identical to the traditional EnKF solution when the ensemble size is $O(100)$. One can question how well the ensemble represents the error statistics for these small ensemble sizes. We may switch from a random sampling interpretation to an error-subspace formulation with such small ensembles.

For those who want to explore the square-root filters further, we refer to Evensen (Evensen, 2009b, Chap. 13) and the overwhelming literature on the ensemble transform Kalman filter (ETKF) and its implementation with localization LETKF (Hunt et al., 2007). For a review comparing different ensemble square-root filters with unified notation, see Vetra-Carvalho et al. (2018).

8.9   Ensemble Subspace Inversion

In Algorithm 5, we have represented the measurement error-covariance matrix by an ensemble of measurement perturbations, $\mathbf{E}$. The inversion is, in this case, computed using an ensemble subspace scheme as was proposed by Evensen (2004), further discussed in Evensen (2009b) and recently in Evensen et al. (2019) and Evensen (2021). This scheme projects the measurement error perturbations onto the ensemble subspace. It computes the pseudo inverse of the following factorization

$$\left(\mathbf{S}\mathbf{S}^{\mathrm{T}} + \mathbf{E}\mathbf{E}^{\mathrm{T}}\right) \tag{8.46}$$

$$\approx \mathbf{S}\mathbf{S}^{\mathrm{T}} + (\mathbf{S}\mathbf{S}^{+})\mathbf{E}\mathbf{E}^{\mathrm{T}}(\mathbf{S}\mathbf{S}^{+})^{\mathrm{T}} \tag{8.47}$$

$$= \mathbf{U}\boldsymbol{\Sigma}\left(\mathbf{I}_N + \boldsymbol{\Sigma}^{+}\mathbf{U}^{\mathrm{T}}\mathbf{E}\mathbf{E}^{\mathrm{T}}\mathbf{U}(\boldsymbol{\Sigma}^{+})^{\mathrm{T}}\right)\boldsymbol{\Sigma}^{\mathrm{T}}\mathbf{U}^{\mathrm{T}} \tag{8.48}$$

$$= \mathbf{U}\boldsymbol{\Sigma}\left(\mathbf{I}_N + \mathbf{Q}\boldsymbol{\Lambda}\mathbf{Q}^{\mathrm{T}}\right)\boldsymbol{\Sigma}^{\mathrm{T}}\mathbf{U}^{\mathrm{T}} \tag{8.49}$$

$$= \mathbf{U}\boldsymbol{\Sigma}\mathbf{Q}\left(\mathbf{I}_N + \boldsymbol{\Lambda}\right)\mathbf{Q}^{\mathrm{T}}\boldsymbol{\Sigma}^{\mathrm{T}}\mathbf{U}^{\mathrm{T}}, \tag{8.50}$$

where we define the singular-value decomposition

$$S = U\Sigma V^T, \tag{8.51}$$

and the identity matrix $I_N \in \Re^{N \times N}$. The eigenvalue decomposition in Eq. (8.49) is of the matrix product in (8.48). Note that this eigenvalue decomposition is most efficiently computed by a singular value decomposition of the product $\Sigma^+ U^T E$. The left singular vectors will then equal the eigenvectors in Q, and the squares of the singular values will equal the eigenvalues in Λ. Thus, the inversion becomes

$$\begin{aligned}
(SS^T + EE^T)^{-1} & \\
\approx (U(\Sigma^\dagger)^T Q)(I_N + \Lambda)^{-1}(U(\Sigma^\dagger)^T Q)^T. & \tag{8.52} \\
= U(\Sigma^\dagger)^T Q(I_N + \Lambda)^{-1} Q^T \Sigma^\dagger U^T &
\end{aligned}$$

The main advantage of this algorithm is that it allows for computing the inverse to a linear cost in the number of measurements, $O(mN^2)$. Also, it is usually easier to simulate measurement perturbations with given statistics than to construct a complete error covariance matrix. The disadvantage is that using a finite ensemble to represent the measurement error covariance matrix introduces additional sampling errors. However, (Evensen, 2021) demonstrated that, by using a larger ensemble to represent E in Eq. (8.46), one could reduce the associated sampling errors to a negligible magnitude and with little additional computational cost. We will demonstrate the consistency of this subspace inversion scheme in the examples in Chap. 13.

8.10   A Note on the EnKF Analysis Equation

Most operational ensemble-based schemes apply an assumption of uncorrelated measurement errors and use a diagonal $C_{dd} = I$, see, e.g., the reviews on data assimilation in the geosciences (Carrassi et al., 2018), weather prediction (Houtekamer & Zhang, 2016), and petroleum applications (Aanonsen et al., 2009). This assumption is employed for simplicity for two reasons. First, the measurement error covariances are often not well known, and additionally, it simplifies the update scheme Eq. (13.5) considerably. With $C_{dd} = I$, Eq. (13.5) becomes

$$Z^a = Z^f + AS^T (SS^T + I)^{-1}(D - HZ), \tag{8.53}$$

which makes it possible to use an efficient algorithm proposed by Hunt et al. (2007) where, by using a Woodbury identity, the EnKF update becomes

$$Z^a = Z^f + A(S^T S + I)^{-1} S^T (D - HZ). \tag{8.54}$$

This modification reduces the size of the matrix to be inverted from $m \times m$ in Eq. (8.53) to $N \times N$ in Eq. (8.54). See also the discussion related to this particular implementation in Evensen et al. (2019, Sect. 3.2).

Alternatively, it is possible to obtain an update equation like Eq. (8.53) if one has access to a factorization $\mathbf{C}_{dd} = \mathbf{C}_{dd}^{\frac{1}{2}}\mathbf{C}_{dd}^{\frac{1}{2}}$ with $\mathbf{C}_{dd}^{\frac{1}{2}}$ being a symmetrical square root of a full rank $\mathbf{C}_{dd}$. E.g., write the eigenvalue decomposition

$$\mathbf{C}_{dd} = \mathbf{Q}\boldsymbol{\Lambda}\mathbf{Q}^{\mathrm{T}} = \mathbf{Q}\boldsymbol{\Lambda}^{\frac{1}{2}}\mathbf{Q}^{\mathrm{T}}\,\mathbf{Q}\boldsymbol{\Lambda}^{\frac{1}{2}}\mathbf{Q}^{\mathrm{T}} = \mathbf{C}_{dd}^{\frac{1}{2}}\mathbf{C}_{dd}^{\frac{1}{2}} \tag{8.55}$$

and define the symmetrical square root

$$\mathbf{C}_{dd}^{\frac{1}{2}} = \mathbf{Q}\boldsymbol{\Lambda}^{\frac{1}{2}}\mathbf{Q}^{\mathrm{T}}, \tag{8.56}$$

and its' inverse

$$\mathbf{C}_{dd}^{-\frac{1}{2}} = \mathbf{Q}\boldsymbol{\Lambda}^{-\frac{1}{2}}\mathbf{Q}^{\mathrm{T}}. \tag{8.57}$$

Now, by scaling the predicted measurement anomalies and the innovations according to

$$\widehat{\mathbf{S}} = \mathbf{C}_{dd}^{-\frac{1}{2}}\mathbf{S}, \tag{8.58}$$

$$\widehat{\mathbf{D}} = \mathbf{C}_{dd}^{-\frac{1}{2}}(\mathbf{D} - \mathbf{HZ}), \tag{8.59}$$

and with some algebra, Eq. (13.5) becomes

$$\mathbf{Z}^{\mathrm{a}} = \mathbf{Z}^{\mathrm{f}} + \mathbf{A}\widehat{\mathbf{S}}^{\mathrm{T}}\left(\widehat{\mathbf{S}}\widehat{\mathbf{S}}^{\mathrm{T}} + \mathbf{I}\right)^{-1}\widehat{\mathbf{D}}, \tag{8.60}$$

and using the Woodbury identity,

$$\mathbf{Z}^{\mathrm{a}} = \mathbf{Z}^{\mathrm{f}} + \mathbf{A}\left(\widehat{\mathbf{S}}^{\mathrm{T}}\widehat{\mathbf{S}} + \mathbf{I}\right)^{-1}\widehat{\mathbf{S}}^{\mathrm{T}}\widehat{\mathbf{D}}. \tag{8.61}$$

Typically, high numerical costs are associated with establishing $\mathbf{C}_{dd}^{\frac{1}{2}}$ and the associated rescalings in Eqs. (8.58) and (8.59), which are both $O(m^2 N)$ operations. Additionally, $\mathbf{C}_{dd}^{\frac{1}{2}}$ needs to be of full rank or a formulation based on pseudo inverses must be employed. Thus, the discussion in this section justifies using the ensemble subspace projection scheme in Eq. (8.52) for computing updates consistently at the cost of $O(mN^2)$, while taking measurement error-correlations into account.

Fully Nonlinear Data Assimilation

9

This chapter provides an introduction to methods that, in theory, samples precisely the posterior pdf. Commonly-used ensemble data-assimilation methods, like the EnKF and EnRML, only sample the posterior pdf correctly in the Gauss-linear case and typically fail in cases with strong nonlinearity. On the other hand, particle methods are also ensemble methods, but they attempt to sample the full posterior pdf, also in problems with multimodal distributions. Their major drawbacks are formed by the convergence issues when the state dimension increases through a phenomenon known as *ensemble degeneracy*. Particle methods work very well for low-dimensional problems, but they require an intelligent implementation with high-dimensional models and affordable ensemble sizes. In the following, we will focus on particle filters and particle flows, which are currently the most promising for highly nonlinear problems in high-dimensions.

9.1 Particle Approximation

The methods discussed in the previous chapters concentrate on finding or approximating certain features of the posterior pdf, such as the mode (typically the variational techniques) or the mean (non-iterative ensemble methods). However, if the posterior pdf is not unimodal and symmetric, then different sampling techniques are needed. Sometimes the pdf can still be described by a small number of parameters, and methods to estimate these parameters can be employed, but this is typically not the case. Often the posterior pdf can have any shape, e.g., being multimodal or heavily skewed. In this case, we can approximate the posterior pdf by using many samples of it.

Let's introduce a new approximation that will significantly ease the computational aspects of the methods discussed in the previous chapter.

© The Author(s) 2022

G. Evensen et al., *Data Assimilation Fundamentals*,
Springer Textbooks in Earth Sciences, Geography and Environment,
https://doi.org/10.1007/978-3-030-96709-3_9

> **Approximation 9** (Particle representation of the pdfs) *It is possible to approximate a probability density function by a finite ensemble of N model states (or particles) as*
>
> $$f(\mathbf{z}) \approx \sum_{j=1}^{N} \frac{1}{N} \delta(\mathbf{z} - \mathbf{z}_j), \tag{9.1}$$
>
> *where $\delta(\cdot)$ denotes the Dirac-delta function.*

We can then use these samples to calculate mean, covariances, higher-order moments, quantiles, etc.

Several data-assimilation methods exist that sample the posterior pdf, and we divide them roughly into sequential Metropolis-like algorithms and parallel particle-filter-like methods. "Sequential" here means that we generate the samples one after the other, and we use the value of the current sample to generate the next one. Examples of this kind of algorithms are Metropolis-Hastings, Gibbs samplers, Langevin samplers, and Hybrid Monte-Carlo. These are known as Markov-Chain Monte-Carlo (MCMC) methods, where the chain part refers to the sequential nature of the sample generation.

The parallel particle-filter-like algorithms generate samples independently from each other or with only small interactions. Examples are standard (bootstrap) particle filters, particle filters with proposal densities, and particle flows. Many of these parallel methods use resampling to ensure sufficient samples to represent the posterior pdf well. Sometimes these particle schemes are also called MCMC methods, but for a different reason. They are connected in physical time via a Markov Chain, for instance, when we propagate the samples in time via a stochastic partial differential equation.

Many schemes use combinations of methods, such as particle-within-Gibbs schemes and particle MCMC. However, strategies that generate samples sequentially cannot be made parallel by construction (although one can run several Markov chains to create different strings of samples). Moreover, we need many samples to converge to a realistic description of the posterior pdf. Thus, it is not common to use these methods in high-dimensional geophysical systems. For that reason, we restrict the discussion to particle filters and flow filters in the following.

As a final note on nonlinear filtering, we touch upon exciting new developments. One is the so-called Schroedinger perspective on data assimilation, in which one tries to draw equal-weight particles from the posterior at time n, based directly on draws from the equal-weight posterior particles at time $n - 1$, see Reich (2019). These methods try to solve the prediction and assimilation problem in one go. No practical algorithms exist for high-dimensional systems, but this is an active research area in the applied mathematics community.

Substantial progress is also reported using coupling methods that try to find an optimal transportation map between prior and posterior pdfs by first finding the map from the prior to a standard Gaussian pdf and then the map from that Gaussian pdf to the posterior pdf. E.g., ElMoselhy and Marzouk (2012) and Spantini et al. (2019)

explore a specific form of the map called the Knothe–Rosenblatt (KR) rearrangement resulting in a triangular map from one pdf to the other. One exciting feature is that many deep learning algorithms with a ReLU activation function have the same structure and are ideal for learning the map. Another exciting part is that for a linear data-assimilation problem, the map is precisely the EnKF.

Another recent development is ensemble-Riemannian data assimilation over the Wasserstein space. The prior and the likelihood are considered marginal pdfs of a coupling pdf, and one calculates the distance between pdfs using the Wasserstein metric. The posterior pdf is then the Wasserstein barycenter of the prior and the likelihood. It becomes similar to the analysis state being the Euclidian barycenter of the prior and likelihood terms in a cost function in linear data assimilation (Tamang et al., 2021). Extensions to high-dimensional problems are described by Tamang et al. (2022), exploring entropic regularization to speed up convergence in the optimization method for the optimal transport map. They applied the technique to a two-layer quasi-geostrophic model.

9.2 Particle Filters

One of the exciting aspects of data assimilation is that we know the solution, and we can often write it down in analytical form, but we do not know how to describe it in practical terms for use in a computer. The expression for the posterior pdf is

$$f(\mathbf{z}|\mathbf{d}) = \frac{f(\mathbf{d}|\mathbf{z})}{f(\mathbf{d})} f(\mathbf{z}), \tag{9.2}$$

which is just Bayes' theorem from Eq. (2.10).

There are several ways to generate samples from the posterior pdf. For instance, similar to ensemble Kalman filters, we can have a set of particles from an ensemble integration of the numerical model. In the ensemble Kalman filter, we assume that these particles describe a Gaussian prior. The particle filter does not apply the Gaussian assumption from Approx. 4, and the prior pdf can have any shape. Thus, the only representation we have of the prior pdf is the set of particles, and in the following, we will describe several methods that explore this feature.

9.2.1 The Standard Particle Filter

We start out by representing the prior using Eq. (9.1), which is the mathematical representation of the prior pdf using samples $\mathbf{z}_i$. Using this representation in the expression for the posterior pdf, we find

$$f(\mathbf{z}|\mathbf{d}) = \sum_{j=1}^{N} w_j \delta(\mathbf{z} - \mathbf{z}_j), \tag{9.3}$$

Algorithm 9 Update scheme with resampling for the standard particle filter

---

1: Input: $\mathbf{Z} \in \mathfrak{R}^{n \times N}$ ▷ Initial model state-vector ensemble
2: Input: $\mathbf{d} \in \mathfrak{R}^{m}$ ▷ Measurements vector
3: **for** $j = 1, N$ **do**   ▷ Log article weights
4:     $\tilde{w}_j = \ln f(\mathbf{d}|\mathbf{z}_j)$
5: **end for**
6: $\tilde{w}_{\max} = \max_j (\tilde{w}_j)$ ▷ Max of log weights
7: **for** $j = 1, N$ **do**   ▷ Unnormalized Particle weights
8:     $\bar{w}_j = \exp(\tilde{w}_{\max} - \tilde{w}_j)$
9: **end for**
10: **for** $j = 1, N$ **do**   ▷ Particle weights
11:     $w_j = \frac{\bar{w}_j}{\sum_i \bar{w}_i}$
12: **end for**
13: **call Resample(w, I)**   ▷ Resampling step
14: **for** $j = 1, N$ **do**   ▷ Resample ensemble
15:     $\mathbf{z}_j = \mathbf{z}_{I_j}$
16: **end for**

---

with the so-called likelihood weights w_j given by

$$w_j = \frac{f(\mathbf{d}|\mathbf{z}_j)}{f(\mathbf{d})} = \frac{f(\mathbf{d}|\mathbf{z}_j)}{\sum_{i=1}^{N} f(\mathbf{d}|\mathbf{z}_i))}. \tag{9.4}$$

Here we have used a standard self normalization to ensure the weights add up to one, which can be derived from $f(\mathbf{d}) = \int f(\mathbf{d}|\mathbf{z}) f(\mathbf{z}) \, d\mathbf{z} \approx \sum_{i=1}^{N} f(\mathbf{d}|\mathbf{z}_i)$, and using the ensemble representation for the prior as above. The full scheme is presented in Algorithm 9.

Within this scheme, we propagate the weighted particles from one assimilation step to the next. At each assimilation step, we assign new weights to the particles. After a few assimilation steps, a particle's weight is proportional to the product of all its previous weights. In practice, this means that the weights of the particles diverge more and more. Typically within a few assimilation steps the relative weight of one particular particle is very close to one, while all the other particles have weights very close to zero. We call the resulting ensemble *degenerate* as it effectively contains just one particle. The weighted ensemble mean equals the particle with near-one weight, and the ensemble variance is close to zero.

One way to avoid this degeneracy problem is to use resampling. If the weights diverge too much, we abandon the low-weight particles and duplicate the high-weight particles such that the total number of particles does not change. A standard measure for particle divergence is the *effective ensemble size*, defined as

$$N_{\text{eff}} = \frac{1}{\sum_{i=1}^{N} w_i^2}, \tag{9.5}$$

where w_i are the normalized weights, such that they add up to one. Typically, re-sampling is introduced when $N_{\text{eff}} \leq 0.8N$.

Algorithm 10 Stochastic Universal Resampling algorithm

1: **subroutine Resample(w, I)**
2: Input: $\mathbf{w} \in \Re^N$
3: $\hat{w}_1 = w_1$
4: **for** $j = 2, N$ **do** ▷ compute cumulative weights
5: $\quad \hat{w}_j = \sum_{i=1}^{j} w_j$
6: **end for**
7: $u \sim \mathcal{U}[0, 1/N]$ ▷ generate a random number
8: $k = 1$
9: **for** $j = 1, N$ **do**
10: $\quad$ **while** $u > \hat{w}_k$ **do**
11: $\quad\quad k = k + 1$
12: $\quad$ **end while**
13: $\quad I_j = k$ ▷ assign an index of the sampled particle
14: $\quad u = u + 1/N$
15: $\quad k = 1$
16: **end for**

Using resampling avoids the accumulation of weights on each particle, and we obtain so-called particle filters with sequential importance resampling (SIR). There are several resampling schemes from which we can choose. The one that leads to the minimum additional random noise is the so-called Stochastic Universal Resampling, which proceeds as denoted in Algorithm 10, see Kitagawa (1996).

When the number of independent observations is large, and large here typically means more than 10, we need 1,00,000 samples or more, even to represent the mean accurately. The reason is that the likelihood will be very peaked if it consists of products of individual likelihoods for each independent observation, which means that the weights will vary enormously over the particles. We often have very many observations in the geosciences, and the SIR method will degenerate even after one assimilation step. In this case, the resampling will result in an ensemble of particles identical to the best member, and the ensemble remains degenerate.

9.2.2 Proposal Densities

Interestingly, we do not have to draw samples from the prior as we can also use samples from another pdf, the so-called proposal pdf $q(\mathbf{z})$. We can rewrite Bayes' theorem from Eq. (9.2) as

$$f(\mathbf{z}|\mathbf{d}) = \frac{f(\mathbf{d}|\mathbf{z})}{f(\mathbf{d})} f(\mathbf{z}) = \frac{f(\mathbf{d}|\mathbf{z})}{f(\mathbf{d})} \frac{f(\mathbf{z})}{q(\mathbf{z})} q(\mathbf{z}), \qquad (9.6)$$

which holds for any $q(\mathbf{z})$ that is nonzero where the prior is nonzero. Assume we have samples from $q(\mathbf{z})$, so we can write

$$q(\mathbf{z}) = \sum_{i=1}^{N} \frac{1}{N} \delta(\mathbf{z} - \mathbf{z}_i), \qquad (9.7)$$

then we find again, using the expression for $q(\mathbf{z})$ Eq. (9.6)

$$f(\mathbf{z}|\mathbf{d}) = \sum_{i=1}^{N} w_i \delta(\mathbf{z} - \mathbf{z}_i), \qquad (9.8)$$

but now with weights

$$w_i = \frac{f(\mathbf{d}|\mathbf{z}_i)}{N \sum_{j=1}^{N} f(\mathbf{d}|\mathbf{z}_j)} \frac{f(\mathbf{z}_i)}{q(\mathbf{z}_i)}. \qquad (9.9)$$

and the weights are now the product of the likelihood weights and the so-called *proposal weights*. It looks like we didn't gain much, but remember that the pdf $q(\mathbf{z})$ can be whatever we choose. For instance, we can make it dependent on the observations, $q(\mathbf{z}|\mathbf{d})$. Thus, we can use particles that already know where the observations are. As an example, we can use posterior samples from an EnKF solution as samples from the proposal density $q(\mathbf{z})$. In that case, all particles will be closer to the observations than samples from the prior. Hence, the likelihood weights will be much closer together. Of course, we need to include the so-called proposal weights $f(\mathbf{z}_j)/q(\mathbf{z}_j)$ in our final expression, but these often have a much smoother distribution than the likelihood weights.

To use this formalism we need to evaluate $f(\mathbf{z}_j)$. In many cases this pdf is assumed to be known, e.g., for pure parameter estimation or for state estimation over just a single time window. In general, we know this pdf for stationary estimation problems.

However, in sequential estimation the posterior pdf at one time step becomes the prior at the next timestep and we do not know the shape of this prior pdf. All we have are a set of particles that represent the prior pdf and it will be impossible to evaluate the value of $f(\mathbf{z}_j)$. Fortunately, we can use the model equations because we know the statistics of the model errors. Thus, we assume we know $f(\mathbf{z}_k|\mathbf{z}_{k-1})$ where k is the time index, and in which $\mathbf{z}$ can be a model state, or parameters when these are estimated sequentially.

For example, let's assume Gaussian additive model errors with mean zero and covariance $\mathbf{C}_{qq}$. In that case we can introduce the transition density

$$f(\mathbf{z}_k|\mathbf{z}_{k-1}) \propto \exp\left(-\frac{1}{2}(\mathbf{z}_k - \mathbf{m}(\mathbf{z}_{k-1}))^{\mathrm{T}} \mathbf{C}_{qq}^{-1}(\mathbf{z}_k - \mathbf{m}(\mathbf{z}_{k-1}))\right). \qquad (9.10)$$

The reason for introducing the transition density of one state to the next is that we can rewrite the prior pdf as

$$f(\mathbf{z}_k) = \int f(\mathbf{z}_k, \mathbf{z}_{k-1}) \, d\mathbf{z}_{k-1} = \int f(\mathbf{z}_k|\mathbf{z}_{k-1}) f(\mathbf{z}_{k-1}) \, d\mathbf{z}_{k-1}. \qquad (9.11)$$

If we now invoke a particle representation at time t_{k-1}, we find the following expression for the prior pdf

$$f(\mathbf{z}_k) = \int f(\mathbf{z}_k|\mathbf{z}_{k-1}) \frac{1}{N} \sum_{j=1}^{N} \delta(\mathbf{z}_{k-1} - \mathbf{z}_{j,k-1}) \, d\mathbf{z}_{k-1} = \sum_{j=1}^{N} \frac{1}{N} f(\mathbf{z}_k|\mathbf{z}_{j,k-1}).$$

$$(9.12)$$

The exciting part of this development is that we have changed from representing the prior by a set of delta functions to using a continuous prior defined by a sum of transition densities. The transition densities are often Gaussian pdfs representing Gaussian model errors *without any further approximation*. Hence, for Gaussian model errors a Gaussian mixture is a natural expression of the prior, where the model error covariance defines the width of the covariance in the Gaussian pdf in each mixture component.

Remember that we want to introduce a proposal density in this formalism to obtain weights with better behavior. With this in mind, we use the particle representation from Eq. (9.12) in Eq. (9.2) to obtain

$$f(\mathbf{z}_k|\mathbf{d}_k) = \frac{f(\mathbf{d}_k|\mathbf{z}_k)}{f(\mathbf{d}_k)} f(\mathbf{z}_k)$$

$$= \frac{f(\mathbf{d}_k|\mathbf{z}_k)}{f(\mathbf{d}_k)} \sum_{j=1}^{N} \frac{1}{N} f(\mathbf{z}_k|\mathbf{z}_{j,k-1}) \qquad (9.13)$$

$$= \frac{f(\mathbf{d}_k|\mathbf{z}_k)}{f(\mathbf{d}_k)} \sum_{j=1}^{N} \frac{1}{N} \frac{f(\mathbf{z}_k|\mathbf{z}_{j,k-1})}{q(\mathbf{z}_k|\mathbf{z}_{k-1}, \mathbf{d})} q(\mathbf{z}_k|\mathbf{z}_{k-1}, \mathbf{d}).$$

Note that we have introduced a transition proposal density $q(\mathbf{z}_k|\mathbf{z}_{k-1}, \mathbf{d})$ that does not only depend on state evolution equations, but that we also allow to depend on the new observations. As before, we multiply and divide the expression in the second line of the equation by q. This division is possible when the support of q is equal to or larger than that of the transition density $f(\mathbf{z}_k|\mathbf{z}_{j,k-1})$. The important element is that we now draw samples from q instead of directly from the model error pdf f. This leads again to

$$f(\mathbf{z}_k|\mathbf{d}_k) = \sum_{i=1}^{N} w_i \delta(\mathbf{z}_k - \mathbf{z}_{i,k}), \qquad (9.14)$$

but now with weights

$$w_i = \frac{f(\mathbf{d}_k|\mathbf{z}_{i,k})}{\sum_{j=1}^{N} f(\mathbf{d}_k|\mathbf{z}_{j,k})} \frac{f(\mathbf{z}_{i,k}|\mathbf{z}_{i,k-1})}{q(\mathbf{z}_{i,k}|\mathbf{z}_{i,k-1}, \mathbf{d}_k)}, \qquad (9.15)$$

where the first part comes from Eq. (9.4). The values of the weights in Eq. (9.15) depend on our choice for q. Since q is a transition density, it is related to state evolution equations. In fact, we can choose any model equation we like to ensure that the weights are less degenerate. The freedom is enormous. Since we can take q to be dependent on the new observations, we can include other data-assimilation methods into a particle filter in a very natural way.

Let's have a look at how to include a stochastic ensemble Kalman filter into a particle filter, as introduced by Papadakis et al. (2010), and as discussed in correction to that scheme in Van Leeuwen (2009). In terms of proposal densities, we can split the stochastic ensemble Kalman filter into two steps, a model evolution step from time t_{k-1} to the observation time t_k, and an update step at time k. We can write for each ensemble member, assuming a linear observation operator and suppressing the superscript f for the forecast

$$
\begin{aligned}
\mathbf{z}_{j,k}^{\mathrm{a}} &= \mathbf{z}_{j,k} + \overline{\mathbf{K}}\big(\mathbf{d}_k - \mathbf{H}\mathbf{z}_{j,k} - \boldsymbol{\epsilon}_j\big) \\
&= \mathbf{m}\big(\mathbf{z}_{j,k-1}\big) + \mathbf{q}_j + \overline{\mathbf{K}}\big(\mathbf{d}_k - \mathbf{H}\mathbf{m}\big(\mathbf{z}_{j,k-1}\big) - \mathbf{H}\mathbf{q}_j - \boldsymbol{\epsilon}_j\big) \qquad (9.16) \\
&= \mathbf{m}\big(\mathbf{z}_{j,k-1}\big) + \overline{\mathbf{K}}\big(\mathbf{d}_k - \mathbf{H}\mathbf{m}\big(\mathbf{z}_{j,k-1}\big)\big) + \big(\mathbf{I} - \overline{\mathbf{K}}\mathbf{H}\big)\mathbf{q}_j - \overline{\mathbf{K}}\boldsymbol{\epsilon}_j,
\end{aligned}
$$

where $\overline{\mathbf{K}}$ is the Kalman gain calculated from the prior ensemble at time k.

We see that the update consists of a deterministic part and a stochastic part. Assuming Gaussian model errors and Gaussian observation errors, the stochastic part is Gaussian with mean zero and covariance

$$
\overline{\boldsymbol{\Sigma}} = \big(\mathbf{I} - \overline{\mathbf{K}}\mathbf{H}\big)\mathbf{C}_{qq}\big(\mathbf{I} - \overline{\mathbf{K}}\mathbf{H}\big)^{\mathrm{T}} + \overline{\mathbf{K}}\mathbf{C}_{dd}\overline{\mathbf{K}}^{\mathrm{T}}. \qquad (9.17)
$$

Hence, the proposal-transition density for each ensemble member of a stochastic ensemble Kalman filter becomes

$$
q\big(\mathbf{z}_k | \mathbf{z}_{j,k-1}, \mathbf{d}\big) = \mathcal{N}\big(\tilde{\mathbf{z}}_j, \overline{\boldsymbol{\Sigma}}\big), \qquad (9.18)
$$

where $\tilde{\mathbf{z}}_j$ results from the deterministic part

$$
\tilde{\mathbf{z}}_j = \mathbf{m}\big(\mathbf{z}_{j,k-1}\big) + \overline{\mathbf{K}}\big(\mathbf{d}_k - \mathbf{H}\mathbf{m}\big(\mathbf{z}_{j,k-1}\big)\big). \qquad (9.19)
$$

Thus, we use EnKF to calculate the "proposed" particles and, after that, obtain their weights from Eq. (9.15). To compute the weights, we evaluate the probability of each EnKF-updated particle from the normal distribution $q(\mathbf{z}_k | \mathbf{z}_{j,k-1}, \mathbf{d})$ in Eq. (9.18). Then we calculate $f(\mathbf{z}_k | \mathbf{z}_{j,k-1})$ for each particle using the Gaussian in Eq. (9.10). The ratio of these two probabilities gives the proposal part of the weights. We multiply this ratio with the likelihood part of the weights from Eq. (9.4) and, after normalization, we have our final weights.

Many other choices are possible too. For instance, one could use a 3DVar on each particle as a proposal. Or we could use the EnKF with reduced observation errors to draw the particles closer to the observations. More extensively, we can also use a 4DVar or an ensemble smoother on each particle (Van Leeuwen et al., 2015). Other suggested proposals include synchronization methods (Pinheiro et al., 2019a, 2019b) and simple nudging schemes (Van Leeuwen, 2010). The point is, one can use every trick in the book and beyond without making any other approximation than Approx. 9. However, the proposed samples should make physical sense and represent the posterior distribution as close as possible at each update step.

9.2.3   The Optimal Proposal Density

One can ask if there is an optimal proposal density, and depending on the definition of optimal, there is. One way to define optimality is to minimize the variance of the weights, leading to the so-called *optimal proposal density*. This density is given by $q(\mathbf{z}_k|\mathbf{z}_{j,k-1}, \mathbf{d}) = f(\mathbf{z}_k|\mathbf{z}_{j,k-1}, \mathbf{d})$. By using the definition of conditional densities as $f(a|b, c) = f(a, b|c)/f(a|c)$ and $f(a, b|c) = f(a|b, c)f(b|c)$, we can write

$$
\begin{aligned}
f\left(\mathbf{z}_k|\mathbf{z}_{j,k-1}, \mathbf{d}_k\right) &= \frac{f\left(\mathbf{z}_k, \mathbf{d}_k|\mathbf{z}_{j,k-1}\right)}{f\left(\mathbf{d}_k|\mathbf{z}_{j,k-1}\right)} = \\
&= \frac{f\left(\mathbf{d}_k|\mathbf{z}_k, \mathbf{z}_{j,k-1}\right)f\left(\mathbf{z}_k|\mathbf{z}_{j,k-1}\right)}{f\left(\mathbf{d}_k|\mathbf{z}_{j,k-1}\right)} = \\
&= \frac{f\left(\mathbf{d}_k|\mathbf{z}_k\right)f\left(\mathbf{z}_k|\mathbf{z}_{j,k-1}\right)}{f\left(\mathbf{d}_k|\mathbf{z}_{j,k-1}\right)},
\end{aligned}
\tag{9.20}
$$

where in the last equality we used that $f(\mathbf{d}_k|\mathbf{z}_k, \mathbf{z}_{j,k-1}) = f(\mathbf{d}_k|\mathbf{z}_k)$ because $\mathbf{d}_k$ does not explicitly depend on $\mathbf{z}_{j,k-1}$ when $\mathbf{z}_{j,k}$ is given. The denominator does not depend on the active variable $\mathbf{z}_k$, and hence is a normalization constant that we do not have to worry about.

We can evaluate the weights of the optimal proposal density without any approximation. From Eq. (9.15) the weights become, using Eq. (9.20),

$$
\begin{aligned}
w_j &= \frac{f(\mathbf{d}_k|\mathbf{z}_{j,k})}{f(\mathbf{d}_k)} \frac{f(\mathbf{z}_{j,k}|\mathbf{z}_{j,k-1})}{f(\mathbf{z}_{j,k}|\mathbf{z}_{j,k-1}, \mathbf{d}_k)} \\
&= \frac{f(\mathbf{d}_k|\mathbf{z}_{j,k})}{f(\mathbf{d}_k)} \frac{f(\mathbf{z}_{j,k}|\mathbf{z}_{j,k-1})}{f(\mathbf{d}_k|\mathbf{z}_{j,k})} \frac{f(\mathbf{d}_k|\mathbf{z}_{j,k-1})}{f(\mathbf{z}_{j,k}, \mathbf{z}_{j,k-1})} \\
&= \frac{f(\mathbf{d}_k|\mathbf{z}_{j,k-1})}{f(\mathbf{d}_k)}.
\end{aligned}
\tag{9.21}
$$

The variance in these optimal proposal weights will be much lower than those of the standard particle filter because of the model error pdf. To see this, we can write

$$
w_j = \frac{f(\mathbf{d}_k|\mathbf{z}_{j,k-1})}{f(\mathbf{d}_k)} = \int \frac{f(\mathbf{d}_k, \mathbf{z}_k|\mathbf{z}_{j,k-1})}{f(\mathbf{d}_k)} \, d\mathbf{z}_k = \int \frac{f(\mathbf{d}_k|\mathbf{z}_k)}{f(\mathbf{d}_k)} f(\mathbf{z}_k|\mathbf{z}_{j,k-1}) \, d\mathbf{z}_k.
\tag{9.22}
$$

Thus, we can write the weights as a convolution of the standard particle filter weights with the model error pdf. Such a convolution always results in a broader pdf as the standard particle filter weights are "smeared out."

When the model and the observation operators are nonlinear, it is not straightforward to generate these optimal proposal particles, i.e., the draws from $f(\mathbf{z}_k|\mathbf{z}_{j,k-1}, \mathbf{d}_k)$. Chorin and Tu (2009), Chorin et al. (2010) and Morzfeld et al. (2012) have developed an efficient scheme named the *implicit particle filter* that partly resolves this problem. For Gaussian observation and model errors, this solution equals a 4DVar estimate for each particle, perturbed by a random error. In each 4Dvar the state covariance at the initial time is zero (and hence, the prior pdf is a delta function centered at that particle). For linear observation operators, we can evaluate this explicitly. The numerator in Eq. (9.20) is a product of two Gaussians, and we know from standard Kalman filters that we can write $f(\mathbf{z}_k|\mathbf{z}_{j,k-1}, \mathbf{d}_k)$ as another Gaussian with mean

$$\tilde{\mathbf{z}}_j = \mathbf{m}(\mathbf{z}_{j,k-1}) + \widetilde{\mathbf{K}}\big(\mathbf{d}_k - \mathbf{Hm}(\mathbf{z}_{j,k-1})\big), \tag{9.23}$$

in which $\widetilde{\mathbf{K}} = \mathbf{C}_{qq}\mathbf{H}^{\mathrm{T}}\big(\mathbf{HC}_{qq}\mathbf{H}^{\mathrm{T}} + \mathbf{C}_{dd}\big)^{-1}$, which is a Kalman gain with model covariance $\mathbf{C}_{qq}$, and with covariance

$$\widetilde{\boldsymbol{\Sigma}} = \big(\mathbf{I} - \widetilde{\mathbf{K}}\mathbf{H}\big)\mathbf{C}_{qq}\big(\mathbf{I} - \widetilde{\mathbf{K}}\mathbf{H}\big)^{\mathrm{T}} + \widetilde{\mathbf{K}}\mathbf{C}_{dd}\widetilde{\mathbf{K}}^{\mathrm{T}}. \tag{9.24}$$

The mean and mode of these transition densities are equal to those of a 4DVar on each particle with prior covariance equal to zero and model error covariance equal to $\mathbf{C}_{qq}$. We do not need the mode, but instead we need to draw from the Gaussian distribution and consider each particle as a weak-constraint 4DVar solution perturbed by a draw from $\mathcal{N}(\mathbf{0}, \widetilde{\boldsymbol{\Sigma}})$. Note the resemblance with using the stochastic EnKF as proposal, as expected. For this particular case, we can generate an analytical expression for the weights as

$$w_j = \frac{f(\mathbf{d}_k|\mathbf{z}_{j,k-1})}{f(\mathbf{d}_k)} = \int \frac{f(\mathbf{d}_k|\mathbf{z}_k)}{f(\mathbf{d}_k)} f(\mathbf{z}_k|\mathbf{z}_{j,k-1})\, d\mathbf{z}_k$$

$$\propto \exp\left(-\frac{1}{2}\big(\mathbf{d}_k - \mathbf{Hm}(\mathbf{z}_{j,k-1})\big)^{\mathrm{T}} \widetilde{C}_{dd}^{-1}\big(\mathbf{d}_k - \mathbf{Hm}(\mathbf{z}_{j,k-1})\big)\right), \tag{9.25}$$

where

$$\widetilde{\mathbf{C}}_{dd} = \mathbf{HC}_{qq}\mathbf{H}^{\mathrm{T}} + \mathbf{C}_{dd}. \tag{9.26}$$

We see that the weight in Eq. (9.25) is the likelihood of particle j starting at the previous time. These weights will be better behaved than the weights of the standard particle filter because we inflate the observation error covariance by a term $\mathbf{HC}_{qq}\mathbf{H}^{\mathrm{T}}$. When the model error is significant, this additional inflation can make the weights much more similar.

9.2.4 Other Particle Filter Schemes

It is easy to show that for high-dimensional systems as encountered in the geosciences the weights are still degenerate when using an EnKF as proposal density. This is even the case with the optimal proposal density with a realistic number of ensemble members, e.g., 100, to compute the proposal ensemble. The community has not yet systematically explored other possibilities of calculating the proposal ensemble, so searching for methods to avoid degeneracy remains an area of active research.

There are mainly two solutions proposed in the literature for avoiding the weights' degeneracy. The first approach uses localization to reduce the number of observations in each local update of the weights. We will discuss localization in more detail in Chap. 10. The second method tries to ensure that all or most particles have equal weight. The reason that we can do better than the optimal proposal density, which minimizes the variance of the weights, is that we sacrifice a few particles to ensure that the rest of the particles have weights that are very similar. Hence, the variance in the weights can be large, but via resampling of the bad particles we can avoid degeneracy of the overall filter. We will not discuss these methods here but we rather refer to e.g., Ades and Van Leeuwen (2013, 2015a, 2015b), Skauvold et al. (2019), Van Leeuwen (2010, 2011), Van Leeuwen and Ades (2013), Zhu et al. (2016), and the review in Van Leeuwen et al. (2019).

A recently proposed third solution is to use methods that avoid particle weights altogether, as discussed in the next section.

9.3 Particle-Flow Filters

In particle flows, one typically starts with equally weighted samples from the prior. Instead of weighing them with the likelihood, as in the standard particle filter, we transform the samples in state space to represent the posterior pdf. This transformation is an iterative process. In the previous chapters, we discussed variational schemes like 4DVar and RML sampling. 4DVar uses an iterative Gauss–Newton method to find the posterior mode, and RML sampling minimizes an ensemble of cost functions to approximately sample the posterior pdf. Particle flow is an iterative ensemble method that in theory correctly samples the posterior pdf.

There is a recent increased interest in methods that dynamically move the particles in state space from equal-weight particles representing the prior, $f(\mathbf{z})$, to equal-weight particles representing the posterior, $f(\mathbf{z}|\mathbf{d})$. In these methods, we seek a potentially stochastic differential equation

$$d\mathbf{z} = \mathbf{m}_s(\mathbf{z})ds + d\mathbf{q}, \qquad (9.27)$$

in artificial time $s \geq 0$ where the deterministic flow map $\mathbf{m}_s$ and the stochastic term $d\mathbf{q}$ define the desired transformation. The stochastic term is drawn from $\mathcal{N}(0, \mathbf{C}_{ff}ds)$, in which $\mathbf{C}_{ff}$ is the covariance matrix of the error in the flow map, i.e., the stochastic forcing. If the initial conditions of the differential equation (9.27) are chosen from a pdf $f_0(\mathbf{z})$ with 0 referring to the initial artificial time, then the solutions follow a distribution characterized by the Fokker–Plank Eq. (2.25) which we write as

$$\frac{\partial f_s}{\partial s} = -\nabla_{\mathbf{z}} \cdot (f_s \mathbf{m}_s) + \nabla_{\mathbf{z}} \cdot \left(\mathbf{C}_{ff} \cdot \nabla_{\mathbf{z}} f_s \right). \qquad (9.28)$$

The initial condition for this equation is $f_0(\mathbf{z}) = f(\mathbf{z})$ and we aim to determine a flow map $\mathbf{m}_s$ and stochastic forcing determined by $\mathbf{C}_{ff}$, such that f_s satisfies the final condition $f_{s_{\text{final}}}(\mathbf{z}) = f(\mathbf{z}|\mathbf{d})$.

9.3.1 Particle Flow Filters via Likelihood Factorization

Several classes of particle-flow filters arise from the likelihood-factorization formalism. To introduce this formulation, let us assume

$$f_s(\mathbf{z}) \propto f(\mathbf{d}|\mathbf{z})^s f(\mathbf{z}), \qquad (9.29)$$

in which $s = 0$ gives us back the prior, and $s = 1$ the posterior pdf. We can take the natural logarithm to find:

$$\ln f_s(\mathbf{z}) \propto s \ln f(\mathbf{d}|\mathbf{z}) + \ln f(\mathbf{z}) + c(s), \qquad (9.30)$$

in which $c(s)$ is a function of the pseudo time s, but not of the state $\mathbf{z}$. If we now take the pseudo-time derivative we find:

$$\frac{1}{f_s}\frac{\partial f_s}{\partial s} = \ln f(\mathbf{d}|\mathbf{z}) + \frac{\partial c(s)}{\partial s}. \qquad (9.31)$$

We now divide the Fokker–Plank Eq. (9.28) by f_s to find:

$$\frac{1}{f_s}\frac{\partial f_s}{\partial s} = -\frac{1}{f_s}\nabla_{\mathbf{z}} \cdot (f_s \mathbf{m}_s) + \frac{1}{f_s}\nabla_{\mathbf{z}} \cdot \left(\mathbf{C}_{f\!f} \cdot \nabla_{\mathbf{z}} f_s\right). \qquad (9.32)$$

Combining the last two equations and taking the gradient to the state $\mathbf{z}$ to eliminate $c(s)$ leads directly to:

$$\nabla \log f(\mathbf{d}|\mathbf{z}) = -\mathbf{m}_s^{\mathrm{T}}\nabla_{\mathbf{z}}^2 \log f_s - \nabla_{\mathbf{z}}(\nabla_{\mathbf{z}} \cdot \mathbf{m}_s) - \nabla_{\mathbf{z}} \log f_s \nabla_{\mathbf{z}} \cdot \mathbf{m}_s + \frac{1}{2}\nabla_{\mathbf{z}}\left(\frac{\nabla_{\mathbf{z}} \cdot (\mathbf{C}_{f\!f}\nabla_{\mathbf{z}} f_s)}{f_s}\right). \qquad (9.33)$$

Thus, we have a nonlinear coupled system of equations whose size is the dimension of the system. However, $\mathbf{m}_s$ has that same dimension, and $\mathbf{C}_{f\!f}$ has that dimension squared, so the number of unknowns is much larger than the number of independent equations. Thus, there are many, in fact, an infinite number of $\mathbf{m}$, $\mathbf{C}_{f\!f}$ combinations that are valid solutions.

Remarkably, and this is truly remarkable, Daum et al. (2018) found an analytical solution of Eq. (9.33) , i.e.,

$$\mathbf{m}_s = -\left(\nabla_{\mathbf{z}}^2 \log f_s\right)^{-1}\left(\nabla_{\mathbf{z}} \log f(\mathbf{d}|\mathbf{z})\right)^{\mathrm{T}}, \qquad (9.34)$$

$$\mathbf{C}_{f\!f} = -\left(\nabla_{\mathbf{z}}^2 \log f_s\right)^{-1}\left(\nabla_{\mathbf{z}}^2 \log f(\mathbf{d}|\mathbf{z})\right)\left(\nabla_{\mathbf{z}}^2 \log f_s\right)^{-1}. \qquad (9.35)$$

This solution's significance is the existence of a closed-form solution for the fully nonlinear data-assimilation problem in terms of the movement of individual particles. Unfortunately, we need the gradient of the logarithm of $f_s(\mathbf{z})$, which is a pdf that we only have an ensemble representation of, so we know it as a sum of Dirac-delta distributions. Hence, this gradient does not exist, and we need to make approximations, e.g., assuming that each particle is not a Dirac-delta distribution but a Gaussian. We have not yet seen this approach explored in any detail for high-dimensional geophysical problems.

In another class of methods, we assume that the stochastic term is zero, and start from a tapering approach, where we gradually increase s such that $s_{\text{final}} = 1$. We now take the limit of increasing number of tapering steps by choosing as steplength $\gamma_i = 1/n_s = \Delta s$ with $\lim_{n_s \to \infty}$, so $\lim_{\gamma_i \to 0}$, or $\lim_{\Delta s \to 0}$, see Daum and Huang (2011, 2013) and Reich (2011). This approach leads to

$$\lim_{\Delta s \to 0} f_{s+\Delta s}(\mathbf{z}) = f_s(\mathbf{z})\left(\frac{f(\mathbf{d}|\mathbf{z})}{f(\mathbf{d})}\right)^{\Delta s}$$

$$= f_s(\mathbf{z})\exp\left(\Delta s\left(\ln f(\mathbf{d}|\mathbf{z}) - \ln f(\mathbf{d})\right)\right) \qquad (9.36)$$

$$\approx f_s(\mathbf{z})\left(1 + \Delta s \ln f(\mathbf{d}|\mathbf{z}) - \Delta s \ln f(\mathbf{d})\right),$$

where we have use a first order Taylor expansion to get to the final line. Hence, we find

$$\frac{\partial f_s(\mathbf{z})}{\partial s} = f_s(\mathbf{z}) \Big(\ln f(\mathbf{d}|\mathbf{z}) - \ln f(\mathbf{d}) \Big)$$
$$= f_s(\mathbf{z}) \Big(\ln f(\mathbf{d}|\mathbf{z}) - c_s \Big), \tag{9.37}$$

with $c_s = \int f_s(\mathbf{z}) \ln f(\mathbf{d}|\mathbf{z}) \, d\mathbf{z}$, which follows directly from integrating the equation over the whole state space, and using $\int f_s(\mathbf{z}) \, d\mathbf{z} = 1$. If we now use the Liouville equation (Jazwinski, 1970) for the evolution of a pdf we can identify

$$\nabla_{\mathbf{z}} (f_s \mathbf{m}_s) = -f_s(\mathbf{z}) \Big(\ln f(\mathbf{d}|\mathbf{z}) - c_s \Big), \tag{9.38}$$

which is an implicit equation for $\mathbf{m}_s$ in terms of f_s. Explicit expressions for $\mathbf{m}_s$ are available for certain pdfs such as Gaussians and Gaussian mixtures (Reich, 2012). These particle-flow filters can be viewed as a continuous limit of the tapering methods, avoiding the need for resampling and jittering. Note that the elliptic partial differential equation (9.38) does not determine $\mathbf{m}_s$ uniquely. Optimal choices in the sense of minimizing the $L_2(f_s)$-norm of $\mathbf{m}_s$ lead to the theory of optimal transportation, see Reich and Cotter (2015) and Villani (2008).

9.3.2 Particle Flows via Distance Minimization

Alternatively, one can define a distance between the intermediate pdf $f_s(\mathbf{z})$ and the posterior pdf, and then find the flow field $\mathbf{m}_s$ that minimizes that distance. Many definitions of the distance between two pdfs exist, and we will use the Kullback–Leibler (KL) divergence here. (The KL divergence is strictly speaking not a distance as it is not symmetric in its arguments, but reducing KL does reduce the distance between the two pdfs.) The following efficient derivation follows Hu and Van Leeuwen (2021).

The KL divergence is given by

$$\mathrm{KL}\big(f_s(\mathbf{z}) \,||\, f(\mathbf{z}|\mathbf{d}) \big) = \int f_s(\mathbf{z}) \ln \frac{f_s(\mathbf{z})}{f(\mathbf{z}|\mathbf{d})} \, d\mathbf{z}, \tag{9.39}$$

and we find the rate of change of the KL divergence with s from

$$\frac{\partial \mathrm{KL}}{\partial s} = \int \frac{\partial f_s(\mathbf{z})}{\partial s} \left(\ln \frac{f_s(\mathbf{z})}{f(\mathbf{z}|\mathbf{d})} - 1 \right) d\mathbf{z}. \tag{9.40}$$

We can rewrite this expression using the Liouville equation for $f_s(\mathbf{z})$ as

$$\frac{\partial \mathrm{KL}}{\partial s} = \int \nabla_{\mathbf{z}} \cdot \big(\mathbf{m}_s(\mathbf{z}) f_s(\mathbf{z}) \big) \left(1 - \ln \frac{f_s(\mathbf{z})}{f(\mathbf{z}|\mathbf{d})} \right) d\mathbf{z}, \tag{9.41}$$

and, using partial integrations twice, we obtain

$$\frac{\partial \mathrm{KL}}{\partial s} = -\int f_s(\mathbf{z}) \big(\mathbf{m}_s(\mathbf{z}) \nabla_{\mathbf{z}} \ln f(\mathbf{z}|\mathbf{d}) + \nabla_{\mathbf{z}} \mathbf{m}_s(\mathbf{z}) \big) d\mathbf{z}. \tag{9.42}$$

Our task now is to find the flow field $\mathbf{m}_s(\mathbf{z})$ that leads to a fast decrease of the KL divergence and thus an efficient mapping from the prior to the posterior pdf. As we have no direct solution to this optimization problem, Liu and Wang (2016) suggest embedding the flow field in a reproducing-kernel Hilbert space (RKHS), such that

$$\mathbf{m}_s(\mathbf{z}) = \left\langle \mathcal{K}(\cdot, \mathbf{z}), \, \mathbf{m}_s(\cdot) \right\rangle, \tag{9.43}$$

in which $\mathcal{K}(\cdot, \mathbf{z})$ is a matrix-valued kernel, so a matrix of functions of two state vectors. Using this result in Eq. (9.42) leads directly to

$$\frac{\partial \mathrm{KL}}{\partial s} = -\left\langle \int f_s(\mathbf{z}) \left[\mathcal{K}(\cdot, \mathbf{z}) \nabla_{\mathbf{z}} \ln f(\mathbf{z}|\mathbf{d}) + \nabla_{\mathbf{z}} \mathcal{K}(\cdot, \mathbf{z}) \right] \, d\mathbf{z}, \, \mathbf{m}_s(\cdot) \right\rangle, \tag{9.44}$$

where we used the linearity of the integral and the inner product to change their order. If we now define $\nabla_{\mathbf{z}} \mathrm{KL}$ as the gradient of the KL distance, i.e., the maximal functional derivative of KL at every state vector $\mathbf{z}$ in the RKHS, we can write the change in KL in the direction of $\mathbf{m}_s$ as

$$\left\langle \nabla_{\mathbf{z}} \mathrm{KL}, \, \mathbf{m}_s \right\rangle. \tag{9.45}$$

By comparing this expression with Eq. (9.44), we can identify

$$\nabla_{\mathbf{z}} \mathrm{KL} = -\int f_s(\mathbf{z}) \left[\mathcal{K}(\cdot, \mathbf{z}) \nabla_{\mathbf{z}} \ln f(\mathbf{z}|\mathbf{d}) + \nabla_{\mathbf{z}} \mathcal{K}(\cdot, \mathbf{z}) \right] \, d\mathbf{z}. \tag{9.46}$$

Hence, by introducing the Reproducing Kernel Hilbert Space, we find an expression for the gradient of KL in terms of an integral that contains the kernel. The critical point is that this gradient is independent of $\mathbf{m}_s$. If we choose the flow field $\mathbf{m}_s$ along this gradient direction

$$\mathbf{m}_s(\mathbf{z}) = -\epsilon \nabla_{\mathbf{z}} \mathrm{KL}(\mathbf{z}), \tag{9.47}$$

where ϵ is a positive scalar, we can use this gradient in a steepest descent minimization of the KL distance. Furthermore, as in variational data-assimilation methods, we can rotate the descent direction to achieve faster convergence. In general, we can use

$$\mathbf{m}_s(\mathbf{z}) = -\mathbf{B} \nabla_{\mathbf{z}} \mathrm{KL}(\mathbf{z}), \tag{9.48}$$

in which $\mathbf{B}$ is a positive definite matrix to our liking. From variational and other iterative methods discussed in Chap. 3, one might want to choose the posterior covariance matrix for $\mathbf{B}$, see, e.g., Eq. (3.17). In practical applications with variables with different physical dimensions, we recommend exploring this freedom of the definition of the matrix $\mathbf{B}$.

Finally, we replace the integral by its empirical approximation by using the particle representation of $f_s(\mathbf{z})$, to obtain

$$\frac{\partial \mathbf{z}_j}{\partial s} = \mathbf{m}_s(\mathbf{z}_j) = \mathbf{B} \frac{1}{N} \sum_{l=1}^{N} \left(\mathcal{K}(\mathbf{z}_j, \mathbf{z}_l) \nabla_{\mathbf{z}_l} \ln f(\mathbf{z}_l|\mathbf{d}) + \nabla_{\mathbf{z}_l} \mathcal{K}(\mathbf{z}_j, \mathbf{z}_l) \right). \tag{9.49}$$

The intuitive explanation of this equation is that the first term in (9.49) pulls the particles towards the mode of the posterior as in a variation method, while the second term acts as a repulsive force that allows for particle diversity. If only the first term is present, the particles will all flow towards the mode of the posterior pdf. As a result, the averaged gradient of the log posterior at each particle, weighted by the kernel, determines the particle flow.

Algorithm 11 Update scheme for a particle flow filter

1: Input: $\mathbf{Z} \in \Re^{n \times N}$ ▷ Initial model state-vector ensemble
2: Input: $\mathbf{d} \in \Re^{m}$ ▷ Measurements vector
3: $\Upsilon = \mathbf{g}(\mathbf{Z})$ ▷ Measure ensemble
4: **call Precondition** $(\mathbf{Z}, \mathbf{B})$ ▷ Preconditioning matrix
5: $\mathbf{m} = 0$
6: **for** $j = 1, N$ **do**
7:   **for** $i = 1, N$ **do**
8:     **call gradLikelihood** $\left(\mathbf{d}, \mathbf{z}_i, \Upsilon_i, \boldsymbol{\phi}_l\right)$ ▷ Alg. 12
9:     **call gradPrior** $\left(\mathbf{Z}, \boldsymbol{\phi}_p\right)$ ▷ Alg. 13
10:     **call kernel** $\left(\mathbf{z}_i, \mathbf{z}_j, \mathcal{K}, \boldsymbol{\phi}_k\right)$ ▷ Alg. 14
11:     $\mathbf{m} = \mathbf{m} + \mathcal{K} \cdot (\boldsymbol{\phi}_l + \boldsymbol{\phi}_p) + \boldsymbol{\phi}_k$
12:   **end for**
13:   $\mathbf{m} = \frac{1}{N}\mathbf{B} \cdot \mathbf{m}$ ▷ Preconditioning
14:   $\mathbf{z}_j = \mathbf{z}_j + \Delta s \mathbf{m}$
15: **end for**

Algorithm 12 Gradient of likelihood for Gaussian observation errors

1: **subroutine gradLikelihood**$(\mathbf{d}, \mathbf{z}_i, \Upsilon_i, \boldsymbol{\phi}_l)$
2: Input: $\mathbf{d} \in \Re^{m}$
3: Input: $\mathbf{z}_i \in \Re^{n}$
4: Input: $\Upsilon_i \in \Re^{m}$
5: $\mathbf{H}_i^T = \nabla_{\mathbf{z}_i} \mathbf{g}(\mathbf{z})$
6: $\boldsymbol{\phi}_l = \mathbf{H}_i^T \mathbf{R}^{-1} (\mathbf{d} - \mathbf{Y}_i)$

Algorithm 13 Gradient of log prior for Gaussian prior

1: **subroutine gradPrior**$(\mathbf{Z}, \boldsymbol{\phi}_p)$
2: Input: $\mathbf{Z} \in \Re^{n \times N}$
3: $\bar{\mathbf{z}} = \sum_{i=1}^{N} \mathbf{z}_i / N$
4: $\boldsymbol{\phi}_p = -\mathbf{C}_{zz}^{-1} (\mathbf{z}_i - \bar{\mathbf{z}})$

The second term avoids this particle collapse by repelling the particles when they become too close. This can easily be seen by choosing a scalar Gaussian kernel, as in Liu and Wang (2016) and Pulido and Van Leeuwen (2019). If we write $\mathcal{K}(\mathbf{z}_j, \mathbf{z}_i) = k(\mathbf{z}_j, \mathbf{z}_i)\mathbf{I}$ and take the gradient to $\mathbf{z}_j$, we obtain

$$\nabla_{\mathbf{z}_l} \mathcal{K}(\mathbf{z}_j, \mathbf{z}_l) \propto (\mathbf{z}_j - \mathbf{z}_l) k(\mathbf{z}_j, \mathbf{z}_l). \tag{9.50}$$

If a component of $\mathbf{z}_j$ is larger than that of $\mathbf{z}_l$, the gradient in Eq. (9.50) is positive, increasing $\mathbf{z}_j$ in that dimension. Thus, the term act as a repelling term. Hu and Van Leeuwen (2021) showed that for sparsely observed systems, a matrix kernel is more efficient than a scalar kernel. The issue with a scalar kernel is that the repelling

---

Algorithm 14 Kernel and its divergence for diagonal Gaussian case

---

1: **subroutine kernel**$(\mathbf{z}_i, \mathbf{z}_j, \mathcal{K}, \boldsymbol{\phi}_k)$

2: Input: $\mathbf{Z} \in \mathfrak{R}^{n \times N}$

3: $\mathcal{K} = 0$

4: **for** $l = 1, n$ **do**

5:      $\mathcal{K}^{(ll)} = \exp\left[-\frac{1}{2} \frac{\left(\mathbf{z}_i^{(l)} - \mathbf{z}_j^{(l)} \right)^2}{\sigma_l^2} \right]$

6:      $\boldsymbol{\phi}_k^{(l)} = -\left(\mathbf{z}_i^{(l)} - \mathbf{z}_j^{(l)} \right) \mathcal{K}^{(ll)} / \sigma_l^2$

7: **end for**

---

term uses the distance between two complete state vectors. The particles converge fast to each other in the space directly influenced by the observations, while that part of the state vector far from the observations shows slow convergence. This slow convergence results in a large distance between particles, hence a tiny repelling force, while the particles collapse in the observed part of the state vector. We can easily avoid this problem by using a simple diagonal kernel with local kernels on the diagonal. We present the particle flow algorithm in Algorithm 11.

Interestingly, Lu et al. (2019) showed that this particle-flow filter converges to the true posterior for any kernel symmetric in its arguments that vanishes at infinity, in the limit of an infinite number of particles. Hence, in that limit the choice of kernel is irrelevant! With a finite number of particles, as in any realistic geophysical application, the choice of the kernel will matter.

Another choice to be made in this scheme is the **B** matrix, which can be seen as a preconditioning matrix for the minimization. By choosing this matrix proportional to a localized ensemble covariance matrix, Hu and Van Leeuwen (2021) demonstrated a practical scheme that works well in problems with hundreds of local modes using only 20 particles. In Chap. 18 we demonstrate the use of a particle-flow implementation with a scalar model and show how the method samples the true posterior distribution, in contrast to traditional assimilation methods, while in Chap. 20 a high-dimensional application to a quasi-geostrophic atmospheric model is described.

Localization and Inflation

10

Localization and inflation have become essential means of mitigating the effects of the low-rank approximation in ensemble methods. Localization increases the effective rank of the ensemble covariance matrix and allows it to fit a large number of independent observations. Thus, we use localization to reduce sampling errors, in combination with inflation, to reduce the underestimation of the ensemble variance caused by the low-rank approximation. These methods are essential for high-dimensional applications, and this chapter will give a general introduction to various formulations of localization and inflation methods.

10.1 Background

The accuracy of data-assimilation methods that exploit a Gaussian prior is strongly dependent on the quality of the prior covariance matrix. This fact is actual both for variational and for ensemble Kalman filters. The prior EnKF ensemble is genuinely a low-rank approximation, and localization is a way to increase the rank of that matrix drastically. That is essential for being able to fit a large number of independent observations. Furthermore, we also use localization to reduce sampling errors, in combination with a technique called inflation, which is essential for high-dimensional applications. The ensemble approximation is not severe when using a sufficiently large ensemble, and we would happily accept the updated ensemble as our posterior estimate. However, we are often limited to using few realizations for computational reasons, limiting the available range of the solution space and leading to significant sampling errors. For example, the Topaz ocean prediction system described by Xie et al. (2017) currently has $\mathcal{O}(10^8)$ state variables and assimilates $\mathcal{O}(10^5)$ measurements in each update step while the ensemble size is 100. Thus, although the "effective" dimension of the model state vector is less than $\mathcal{O}(10^8)$, the state space is vastly

© The Author(s) 2022

G. Evensen et al., *Data Assimilation Fundamentals*,
Springer Textbooks in Earth Sciences, Geography and Environment,
https://doi.org/10.1007/978-3-030-96709-3_10

undersampled, and the ensemble size is too small to represent all the information provided by a large number of measurements. We would not be able to run these high-dimensional systems without localization and inflation.

Suppose the measurements contain spatial variability on scales that the model cannot simulate. In that case, localization allows introducing these finer scales in the updated ensemble and obtaining a model that better fits the observations. It then becomes a partly philosophical and partly computational question whether it makes sense to introduce these fine scales in the model or not. The alternative would be to treat the finer scales in the measurements as representation errors. We can then reduce the measurements' impact by increasing their error variance and specifying correlated measurement errors. Still, we need localization in large-scale systems since the ensemble space cannot generally accommodate all the information provided by the measurements.

We also use localization in particle filters, but for a different reason. There the issue is filter degeneracy and weight collapse due to a too large number of independent observations. Localization limits the number of measurements each gridpoint sees, reducing the degeneracy problem. This chapter provides the basis for different localization methods and formulations, including Kalman gain localization, covariance localization, and local analysis. We will see that localization and inflation are well developed for the EnKF, while it remains a severe obstacle in particle filtering.

10.2   Various Forms of the EnKF Update

This section contains various forms of the EnKF update equations that we will use throughout this chapter. With reference to definitions given in Chap. 8, we can write the EnKF update Eq. (8.29), in the following alternative but equivalent forms:

$$\mathbf{Z}^a = \mathbf{Z}^f + \overline{\mathbf{C}}_{zy}\left(\overline{\mathbf{C}}_{yy} + \overline{\mathbf{C}}_{dd}\right)^{-1}\left(\mathbf{D} - \mathbf{g}(\mathbf{Z}^f)\right) \tag{10.1}$$

$$= \mathbf{Z}^f + \mathbf{A}\mathbf{Y}^T\left(\mathbf{Y}\mathbf{Y}^T + \mathbf{E}\mathbf{E}^T\right)^{-1}\left(\mathbf{D} - \mathbf{g}(\mathbf{Z}^f)\right) \tag{10.2}$$

$$= \mathbf{Z}^f + \overline{\mathbf{K}}\left(\mathbf{D} - \mathbf{g}(\mathbf{Z}^f)\right) \tag{10.3}$$

$$= \mathbf{Z}^f + \overline{\mathbf{C}}_{zy}\mathbf{B} \tag{10.4}$$

$$= \mathbf{Z}^f + \mathbf{A}\mathbf{W} \tag{10.5}$$

$$= \mathbf{Z}^f\overline{\mathbf{T}}. \tag{10.6}$$

Here each column of $\mathbf{Z}$ represents an ensemble realization, and the corresponding column in $\mathbf{D} - \mathbf{g}(\mathbf{Z}^f)$ is this realization's innovation vector, i.e., the difference between perturbed and predicted observations. The various representations of the EnKF update allow for making different interpretations of it's ensemble low-rank approximations. We eliminate apparence of the projection $\mathbf{A}^\dagger\mathbf{A}$ from the analysis equations by using Eq. (8.7) in the following discussion.

In Eqs. (10.2) and (10.5) we have used Eq. (8.32).

In Eq. (10.3) we have defined the ensemble representation of the Kalman gain from Eq. (6.39) as

$$\overline{\mathbf{K}} = \overline{\mathbf{C}}_{zy}\left(\overline{\mathbf{C}}_{yy} + \overline{\mathbf{C}}_{dd}\right)^{-1},   \tag{10.7}$$

which leads to an interpretation of the analysis as computing the update as linear combinations of the m columns of the Kalman gain matrix.

The representer update in Eq. (10.4) is the ensemble version of the formulation in Eqs. (6.40)–(6.42). We can interpret the representer update in Eq. (10.4) as computing the update by adding a linear combination of covariance functions (or representer functions), one for each measurement, to the prior. Thus, $\overline{\mathbf{C}}_{zy}$ is the representer functions' ensemble representation and the columns in $\mathbf{B}$ define the linear combinations of representer functions used to create the analyzed ensemble members. The matrix $\mathbf{B}$ is the solution of the following m-dimensional linear system of equations with N right-hand sides,

$$\left(\mathbf{YY}^{\mathrm{T}} + \mathbf{EE}^{\mathrm{T}}\right)\mathbf{B} = \mathbf{D} - \mathbf{g}\left(\mathbf{Z}^{\mathrm{f}}\right).   \tag{10.8}$$

Thus, both the representer formulation and the Kalman gain version of the analysis update have a similar interpretation. Both Eqs. (10.3) and (10.4) computes the solution in observation space as defined by the dimension of the matrix $\left(\mathbf{YY}^{\mathrm{T}} + \mathbf{EE}^{\mathrm{T}}\right)$. However, in the case with $N < m$ the update is still of low rank and we can compute it more efficiently using Eqs. (10.5) or (10.6).

10.3   Impact of Sampling Errors in the EnKF Update

There are three major consequences of using a low-rank ensemble approximation when computing the analysis updates and we will discuss each of them in this section.

10.3.1   Spurious Correlations

Initially, if approaching the ensemble methods with the perspective of the Kalman filtering community, the first obvious consequence of using an ensemble of limited size is a poor representation of the covariance functions and the error covariance matrix. And, when using a finite ensemble size, we introduce long-range spurious or unphysical correlations in the covariances $\overline{\mathbf{C}}_{zy}$. Thus, a measurement may influence the update throughout the model domain due to the spurious correlations. Another consequence of the spurious correlations is that they lead to an unrealistic reduction of the ensemble variance far from the measurements' locations, leading to under-estimated prediction uncertainty and possible filter divergence. This observation led to the introduction of methods for covariance localization, as we discuss below.

10.3.2 Update Confined to Ensemble Subspace

Evensen (2003) explicitly showed that the EnKF update equation computes the analyzed ensemble realizations as linear combinations of the prior ensemble realizations
(see Eq. 10.6) where we define the transition matrix using notation from Chap. 8 and
Evensen et al. (2019)

$$\mathbf{Z}^a = \mathbf{Z}^f + \mathbf{A}\mathbf{W} \tag{10.9}$$

$$= \mathbf{Z}^f + \mathbf{Z}^f\left(\mathbf{I} - \frac{1}{N}\mathbf{1}\mathbf{1}^T\right)\mathbf{W}/\sqrt{N-1} \tag{10.10}$$

$$= \mathbf{Z}^f + \mathbf{Z}^f\mathbf{W}/\sqrt{N-1} \tag{10.11}$$

$$= \mathbf{Z}^f\left(\mathbf{I} + \mathbf{W}/\sqrt{N-1}\right) \tag{10.12}$$

$$= \mathbf{Z}^f\mathbf{T}, \tag{10.13}$$

with $\mathbf{W}$ defined from Eq. (8.33) as

$$\mathbf{W} = \mathbf{Y}^T\left(\mathbf{Y}\mathbf{Y}^T + \mathbf{E}\mathbf{E}^T\right)^{-1}\left(\mathbf{D} - \mathbf{g}(\mathbf{Z}^f)\right), \tag{10.14}$$

and using $\mathbf{1}^T\mathbf{W} = 0$ (Evensen et al., 2019). Thus, using the EnKF update equation it
is impossible to obtain an update outside the subspace spanned by the prior ensemble.
As in the Topaz system referenced above, the ensemble space may be too restricted
to allow for a realistic update incorporating all the measurements' information. This
issue is different from spurious correlations and is directly related to the low rank of
the prior ensemble.

10.3.3 Ensemble Representation of the Measurement Information

Another issue with the EnKF update equation is that it effectively projects the measurements onto the ensemble subspace. Let's use a measurement error covariance
matrix $\mathbf{C}_{dd}$ of full-rank and use the Woodbury corollary from Eq. (6.10) to obtain

$$\mathbf{Z}^a = \mathbf{Z}^f + \mathbf{A}\mathbf{Y}^T\left(\mathbf{Y}\mathbf{Y}^T + \mathbf{C}_{dd}\right)^{-1}\left(\mathbf{D} - \mathbf{g}(\mathbf{Z}^f)\right). \tag{10.15}$$

$$= \mathbf{Z}^f + \mathbf{A}\left(\mathbf{I}_N + \mathbf{Y}^T\mathbf{C}_{dd}^{-1}\mathbf{Y}\right)^{-1}\mathbf{Y}^T\mathbf{C}_{dd}^{-1}\mathbf{D}'. \tag{10.16}$$

$$= \mathbf{Z}^f + \mathbf{A}\left(\mathbf{I}_N + \widetilde{\mathbf{Y}}^T\widetilde{\mathbf{Y}}\right)^{-1}\widetilde{\mathbf{Y}}^T\widetilde{\mathbf{D}}', \tag{10.17}$$

where we have defined $\widetilde{\mathbf{Y}} = \mathbf{C}_{dd}^{-\frac{1}{2}}\mathbf{Y}$ and $\widetilde{\mathbf{D}}' = \mathbf{C}_{dd}^{-\frac{1}{2}}\mathbf{D}'$ which is a normalization or
scaling of $\mathbf{Y}$ and $\mathbf{D}'$ by the inverse square root of the measurement error covariance
matrix. The important conclusion is that the EnKF update effectively projects the
scaled innovations onto the ensemble subspace through the multiplication $\widetilde{\mathbf{Y}}^T\widetilde{\mathbf{D}}'$.
Thus, the EnKF update removes all the information in the measurements that the
ensemble of predicted measurements cannot represent. This issue is also directly
related to the low rank of the prior ensemble.

10.4 Localization in Ensemble Kalman Filters

There are different ways to reduce the impact of sampling errors and the low rank of the prior covariance matrix in ensemble methods. A common approach is to damp the long-range spurious correlations, and *covariance localization* (Hamill et al., 2001; Houtekamer & Mitchell, 2001) is one such method. Another alternative is to use the *local analysis* first used by Haugen and Evensen (2002) and later explained in more detail by Evensen (2003), where one updates variables on subsets of gridpoints using only the nearby observations that we know should impact these variables. In both methodologies, we restrict the influence radius of observations, effectively decoupling regions of the state space far apart. The local analysis allows for different linear combinations of prior ensemble members to be used in the distinct parts of the state space, effectively increasing the prior ensemble-covariance matrix's rank by orders of magnitude. In a review paper, Sakov and Bertino (2011) discussed the formal similarities between covariance localization and local analysis and concluded that in practice, the two approaches should yield somewhat similar results, and one should base the choice of localization method on criteria such as computational efficiency. We refer to Sakov and Bertino (2011) for an overview of early papers discussing various localization methods, while the more recent review of Chen and Oliver (2017) analyzes different localization scheme when used with an iterative ensemble smoother.

10.4.1 Covariance Localization

In covariance localization, (Anderson, 2003; Bishopetal., 2001; Hamilletal., 2001; Houtekamer & Mitchell, 2001; Whitaker & Hamill, 2002) we use a damping operator that eliminates long-range spurious correlations in the state covariance matrix $\overline{\mathbf{C}}_{zz}$. Typically, we would damp each covariance function by multiplying it with a damping function that equals one at the diagonal element and zero at elements corresponding to variables far from the diagonal element. It is common to write the update equations with covariance localization as

$$\mathbf{Z}^{\mathrm{a}} = \mathbf{Z}^{\mathrm{f}} + \left(\boldsymbol{\rho}_{n \times n} \circ \overline{\mathbf{C}}_{zz}\right)\mathbf{H}^{\mathrm{T}}\left(\mathbf{H}\left(\boldsymbol{\rho}_{n \times n} \circ \overline{\mathbf{C}}_{xx}\right)\mathbf{H}^{\mathrm{T}} + \overline{\mathbf{C}}_{dd}\right)^{-1}\left(\mathbf{D} - \mathbf{g}(\mathbf{Z}^{\mathrm{f}})\right). \quad (10.18)$$

Here we have introduced the Schur (or Hadamard) product denoted by $\circ$ of element-wise multiplication of two matrices. The matrix of damping functions, $\boldsymbol{\rho}_{n \times n}$, acts on the covariance functions in $\overline{\mathbf{C}}_{zz}$, and uses a scaling equal to one near a measurement and then gradually reduces to zero further away from the measurement location. A commonly used damping function is the one by Gaspari and Cohn (1999) but see also Furrer and Bengtsson (2007) for an empirical formula that also accounts for the ensemble size. Covariance localization requires us to compute the full state error covariance matrix, which is an overwhelming task for large systems. Since the damping matrix is typically full rank, the Schur product will also be of full rank.

Chen and Oliver (2017) pointed out that with $\mathbf{H}$ being a local and linear measurement operator (rows of $\mathbf{H}$ contains zeros and one element equal to one that corresponds to the measurement), the following applies

$$\left(\rho_{n\times n}\circ\overline{\mathbf{C}}_{xx}\right)\mathbf{H}^{\mathrm{T}}=\left(\rho_{n\times n}\mathbf{H}^{\mathrm{T}}\right)\circ\left(\overline{\mathbf{C}}_{xx}\mathbf{H}^{\mathrm{T}}\right). \tag{10.19}$$

$$\mathbf{H}\left(\rho_{n\times n}\circ\overline{\mathbf{C}}_{xx}\right)\mathbf{H}^{\mathrm{T}}=\left(\mathbf{H}\rho_{n\times n}\mathbf{H}^{\mathrm{T}}\right)\circ\left(\mathbf{H}\overline{\mathbf{C}}_{xx}\mathbf{H}^{\mathrm{T}}\right), \tag{10.20}$$

Thus, it is possible to replace the Schur product with the $n \times n$-dimensional state covariance matrix, by a Schur product with a $n \times m$ matrix. Even more importantly, we do not need to form the full state covariance matrix as it suffices to form the covariance matrix between the state variables and the predicted measurements. This observation leads to methods for localization in the observation space.

10.4.2   Localization in Observation Space

For computational efficiency, Houtekamer and Mitchell (2001) proposed to approximate the localization in Eq. (10.18) by writing Eq. (10.1) as

$$\mathbf{Z}^{\mathrm{a}}=\mathbf{Z}^{\mathrm{f}}+\rho_{n\times m}\circ\overline{\mathbf{C}}_{zy}\left(\rho_{m\times m}\circ\overline{\mathbf{C}}_{yy}+\overline{\mathbf{C}}_{dd}\right)^{-1}\left(\mathbf{D}-\mathbf{g}(\mathbf{Z}^{\mathrm{f}})\right). \tag{10.21}$$

We have defined $\overline{\mathbf{C}}_{zy}=\overline{\mathbf{C}}_{zz}\mathbf{H}^{\mathrm{T}}$ and $\overline{\mathbf{C}}_{yy}=\mathbf{H}\overline{\mathbf{C}}_{zz}\mathbf{H}^{\mathrm{T}}$, with $\mathbf{H}$ being a linear measurement operator. With nonlinear measurement operators we can represent the ensemble covariances by their ensemble approximations

$$\overline{\mathbf{C}}_{zy}=\mathbf{A}\mathbf{Y}^{\mathrm{T}}, \tag{10.22}$$

$$\overline{\mathbf{C}}_{yy}=\mathbf{Y}\mathbf{Y}^{\mathrm{T}}. \tag{10.23}$$

using the definition in Eqs. (8.3) and (8.8). As pointed out by Chen and Oliver (2017), the relations in Eqs. (10.19) and (10.20) are not valid for general observation operators. Thus, we need to define $\rho_{n\times m}$ and $\rho_{m\times m}$ according to the problem at hand.

Chen and Oliver (2017) also discussed *Kalman-gain localization* where one localizes the Kalman-Gain matrix directly as

$$\mathbf{Z}^{\mathrm{a}}=\mathbf{Z}^{\mathrm{f}}+\rho_{n\times m}\circ\mathbf{K}\left(\mathbf{D}-\mathbf{g}(\mathbf{Z}^{\mathrm{f}})\right). \tag{10.24}$$

Similarly to Eq. (10.25) we must compute the full $\overline{\mathbf{C}}_{zy}\in\mathfrak{R}^{n\times m}$, but we ignore the localization of $\overline{\mathbf{C}}_{yy}$. Kalman-gain localizaion is popular in the petrolium community and does indeed reduce the impact of spurious correlations. However, it does not remove the spurious correlations in $\mathbf{C}_{yy}$ that induce unphysical dependencies between remote meaurements and thereby reduces their impact on the update.

10.4.3 Localization in Ensemble Space

When using Eqs. (10.22) and (10.23) we can write Eq. (10.21) as

$$\mathbf{Z}^{a} = \mathbf{Z}^{f} + \rho_{n \times m} \circ \left(\mathbf{A}\mathbf{Y}^{T}\right)\left(\rho_{m \times m} \circ \left(\mathbf{Y}\mathbf{Y}^{T}\right) + \mathbf{E}\mathbf{E}^{T}\right)^{-1}\left(\mathbf{D} - \mathbf{g}(\mathbf{Z}^{f})\right). \quad (10.25)$$

Thus, we can extend this discussion to search for a localization approach in the ensemble subspace where the Schur product acts directly on the ensemble anomalies. As for the Kalman-gain localization, we neglect the localization of the covariance of the predicted measurement anomalies. Thus, we write Eq. (10.25) as

$$\mathbf{Z}^{a} = \mathbf{Z}^{f} + \rho_{n \times N} \circ \mathbf{A}\left\{\mathbf{Y}^{T}\left(\mathbf{Y}\mathbf{Y}^{T} + \mathbf{E}\mathbf{E}^{T}\right)^{-1}\left(\mathbf{D} - \mathbf{g}(\mathbf{Z}^{f})\right)\right\}. \quad (10.26)$$

However, in this formulation, we need to redefine our interpretation of the data and represent them in the ensemble subspace. Following the derivation in Eqs. (10.15)–(10.17) we have

$$\mathbf{Z}^{a} = \mathbf{Z}^{f} + \rho_{n \times N} \circ \mathbf{A}\left\{\mathbf{Y}^{T}\left(\mathbf{Y}\mathbf{Y}^{T} + \mathbf{C}_{dd}\right)^{-1}\left(\mathbf{D} - \mathbf{g}(\mathbf{Z}^{f})\right)\right\}. \quad (10.27)$$

$$= \mathbf{Z}^{f} + \rho_{n \times N} \circ \mathbf{A}\left\{\left(\mathbf{I}_{N} + \widetilde{\mathbf{Y}}^{T}\widetilde{\mathbf{Y}}\right)^{-1}\widetilde{\mathbf{Y}}^{T}\widetilde{\mathbf{D}}'\right\}, \quad (10.28)$$

with $\widetilde{\mathbf{Y}} = \mathbf{C}_{dd}^{-\frac{1}{2}}\mathbf{Y}$ and $\widetilde{\mathbf{D}}' = \mathbf{C}_{dd}^{-\frac{1}{2}}\mathbf{D}'$ as before. Thus, by projecting the measurement innovations onto the ensemble subspace through the product $\widetilde{\mathbf{Y}}^{T}\widetilde{\mathbf{D}}'$, we approximately represent the measurement's information by N projected measurements. Note that it is not possible to use a physical distance in the localization scheme for the N projected measurements. However, the adaptive localization schemes discussed below might be used in this case. The main problem with this approach is the following: one reason for applying localization is that the measurements contain more information than the ensemble subspace can accommodate. Thus, localization allows us to compute a more prosperous update that is not confined to the ensemble subspace. But, from Eq. (10.28), we project the original measurements onto the ensemble subspace, and we thereby lose all measured information that the ensemble subspace cannot represent. So, in this case, why would we localize at all? But, see also the ideas suggested by Buehner (2005).

10.4.4 Local Analysis

Brusdal et al. (2003) and Haugen and Evensen (2002) used a distance-based local-analysis scheme in an ocean circulation model, where for each vertical column of gridpoints, they updated the state variables using only nearby-located measurements, we quote from Haugen and Evensen (2002) "Note that only data located at gridpoints within a certain influence radius (here chosen to 40 km) are used in the update of the state variables in each gridpoint. This is a common procedure normally denoted as a local analysis." Evensen (2003) gave a more detailed explanation of the local analysis, and later, Ott et al. (2004) introduced the popular local ensemble transform Kalman filter (LETKF) using the same local-analysis concept.

In local analysis, we first need to select subsets of variables to update independently. In principle, we could update the elements in the state vector one by one, but that will generally become too computationally expensive. So instead, a sensible approach is to select all variables associated with a vertical column of gridpoints. Alternatively, if vertical localization is essential, we can split the model grid into subgroups of grid points with a limited horizontal and vertical extent. After that, we must select which measurements to include in the analysis update for each subgroup. We can often use a distance-based approach to retain all the measurements located within a specific range from the subgroup to be updated. However, in some applications, we have so-called non-local measurements where the measured information results from non-local physical processes that can extend over large parts of the model domain. One such example includes pressure transients between wells in reservoir models. Adaptive localization may be a better alternative in these cases. We then select measurements based on their correlation with a group of variables being significant (Neto et al., 2021).

When using local analysis, we can write the update equation Eq. (10.5) in the three following forms

$$\mathbf{Z}_l^a = \mathbf{Z}_l^f + \mathbf{A}_l \mathbf{W}_l \tag{10.29}$$

$$= \mathbf{Z}_l^f + \overline{\mathbf{K}}_l\big(\mathbf{D}_l - \mathbf{g}_l(\mathbf{Z}^f)\big), \tag{10.30}$$

$$= \mathbf{Z}_l^f + \mathbf{A}_l\Big\{\big(\mathbf{I}_N + \mathbf{Y}_l^T \mathbf{C}_{dd,l}^{-1} \mathbf{Y}_l\big)^{-1} \mathbf{Y}_l^T \mathbf{C}_{dd,l}^{-1} \mathbf{D}_l'\Big\}, \tag{10.31}$$

where l runs over the different subgroups of local model variables. Here $\mathbf{W}_l$ and $\overline{\mathbf{K}}_l$ are local variants of Eqs. (10.14) and (10.7) evaluated using the selected observations for each l. As each local update uses an individual Kalman gain or weight matrix, the updated ensemble will no longer be confined to the prior ensemble space. Note that the local analyses are straightforward to parallelize as the computations for different values of l are independent.

There is a computational advantage of using local analysis rather than covariance (or Kalman gain) localization when working with ensemble methods. The reason is that the Kalman-gain matrix is of low rank unless the number of measurements is less than the number of ensemble realizations. Thus, forming the Kalman gain matrix and performing a Schur product is significantly more computationally demanding than computing the local analysis. This statement is true even though it is possible to calculate the Kalman gain matrix and the corresponding update row by row in parallel (Chen & Oliver, 2017). However, in the local analysis it is not uncommon to have $n \sim m \sim N$, and when the ratio m/N is sufficiently small, the Kalman gain update in Eq. (10.30) becomes more computational efficient than the update in Eq. (10.29).

A variant of the local-analysis scheme combines the formulas in Eq. (10.30) with the tapering used for covariance or Kalman gain localization (Chen & Oliver, 2017). This approach was also applied by Neto et al. (2021). Alternatively, the local analysis with observation taper (Greybush et al., 2011; Hunt et al., 2007) uses the form Eq. (10.31) when computing the local updates and then tapers the (in their case diagonal) inverse of the error covariance matrix $\mathbf{C}_{dd}^{-1}$ for each local update. When using the local analysis in the form Eq. (10.29), we can inflate the variance of the remotest located measurements by scaling selected rows in $\mathbf{E}$ to obtain the same effect. The tapering of the local updates reduces the impact of the local measurements located furthest from the gridpoints being updated and reduces the discontinuities in the updated solution. Note that covariance tapering of the local analysis updates is affordable since both the local state dimension and the number of measurements are very low compared to the global analysis update.

10.5   Adaptive Localization

In cases with non-local measurements, where it is impossible to use distance-based tapering, it might be possible to use an adaptive localization method. In adaptive localization, we use the ensemble correlations between a predicted measurement and the state variables at a particular gridpoint to determine if we should update these variables using this measurement. The most straightforward approach is to truncate all observations that have correlations below a certain level. This approach eliminates the impact of spurious correlations but also removes weak but physical correlations. To improve on this approach, Anderson (2007b) proposed using many small ensembles to check if correlations are significant, while Bishop and Hodyss (2007) uses the correlation function to derive a tapering function. Fertig et al. (2007) used the ensemble correlation function to decide whether to update a variable or not. Evensen (2009b, Chap. 15) discussed adaptive localization based on the truncation of small covariances in an example with an advection equation. He found that the approach worked well but led to small discontinuities in the updates, which will likely cause problems in many nonlinear models. Luo et al. (2019) and Luo and Bhakta (2020) have continued this work and developed tuned schemes where they combine truncation-based adaptive localization with tapering of each local update. Neto et al. (2021) and Soares et al. (2021) used this approach successfully in petroleum applications. It is essential to focus the localization issue of non-local observations, not on the prior covariance itself, but the state-observation correlations. As Van Leeuwen (2019) shows, non-local measurements can influence distant state variables, not physically connected in the prior covariance, but they become connected via the observation operator.

10.6   Localization in Time

Localization also plays an important role when using iterative ensemble smoothers like EnRML and ESMDA. Particularly when we use iterative ensemble smoothers in sequential data assimilation, the accumulation of errors resulting from spurious correlations will impact the results significantly. It becomes more tricky to define which observations should impact which state variables when we localize in time. For instance, we know that the information propagates on the model characteristics for hyperbolic models, such as the linear advection equation. In this case, when updating the solution at a gridpoint, we should include all measurements located close to the characteristic line intersecting this grid point. For more complex nonlinear models, the situation becomes even more complicated.

On the other hand, many realistic models have chaotic behavior, limiting the time interval over which measurements will impact the update. There are several proposed solutions. The simplest is to use a larger influence radius with time localization to include all relevant observations as, e.g., used in Brusdal et al. (2003). Bocquet (2015) discusses several time localization schemes where the localization domain effectively propagates with the dynamical flow. Amezcua et al. (2017) show how a weak-constraint ensemble smoother strongly reduces the severity of the issue because the model errors can absorb observation influences local in time and transfer them to the state variables that propagate through time. But maybe adaptive localization will be even more helpful in the case of localization in time.

10.7 Inflation

Anderson and Anderson (1999) suggested using an approach named inflation to counteract the excessive variance reduction caused by spurious correlations in the update. Inflation is generally needed to avoid filter divergence in operational ensemble data-assimilation systems with small ensemble sizes. We can implement inflation as a scaling of the ensemble anomalies

$$\mathbf{A} \leftarrow \rho \mathbf{A}, \tag{10.32}$$

where ρ is a factor slightly larger than one. Today, most operational ensemble data-assimilation systems apply some calibrated inflation to counteract different error sources. We can inflate before or after the analysis update. If applied to the forecast ensemble, inflation is a way to account for model errors and compensate for the low-rank approximation, increasing the predicted ensemble variance. If we inflate the analysis ensemble, inflation accounts for errors introduced by the analysis scheme, e.g., spurious correlations, the approximate representation of the measurement error covariance matrix, and possibly adverse localization effects. The standard procedure is to inflate the analysis update and calibrate the inflation factor to obtain a data-assimilation system in good agreement with observations. As such, inflation is an approach that tries to correct "everything" wrong in the system.

Some papers attempt to estimate an optimal inflation parameter adaptively. Anderson (2009) proposed a method for adaptively estimating a spatially and temporally varying inflation parameter using a Bayesian algorithm. The algorithm is recursive and updates the inflation parameter with time. Wang and Bishop (2003) uses the sequence of innovation statistics to compute the covariance inflation, while Anderson (2007a) estimates the inflation parameter as part of the state vector. Sacher and

Bartello (2008) discuss the sampling errors in EnKF and proposes an analytical expression for the optimal covariance inflation method, which depends on the Kalman gain, the analyzed variance, and the number of realizations. But see also the adaptive inflation estimation by Evensen (2009b) targeting the impact of spurious correlations.

10.8 Localization in Particle Filters

In particle filters, we introduce localization for a different reason than in ensemble Kalman filter methods. As we have seen in Chap. 9, particle filters do not rely on accurate estimation of covariance matrices, which is essential for the success of variational methods and (iterative) ensemble Kalman filters. The problem here is that the weights are degenerate when the number of independent observations is large, see, e.g., Ades and Van Leeuwen (2015a), and Snyder et al. (2008, 2015). And "large" is minor for geophysical applications, where more than ten independent observations typically force us to use tens of thousands of particles. Hence, we use localization in particle filters to reduce the number of measurements in the likelihood of each gridpoint.

The idea of using localization in particle filters was first introduced in 2003 in three papers (Bengtsson et al., 2003; Van Leeuwen 2003a, b). Localization in particle filters faces two main problems. One issue with localization is that different sets of particles will survive after resampling different gridpoints. It is hard to connect these different particle sets from different gridpoints to form smooth global particles that the model equations can propagate. Practical solutions all diminish the influence of the observations, e.g., by setting a minimum weight for each particle, which restricts the size of the update of the prior particles (Poterjoy, 2016; Poterjoy & Anderson, 2016), or reducing the observation space to that of the ensemble space of the prior particles before calculating weights Potthast et al. (2019), or combinations of these. Typically, further smoothing is needed, such as relaxation to prior particles.

Another localization issue is that in some geophysical systems, such as the atmosphere, the number of measurements inside the localization radius will still be too large, and the filter becomes degenerate. The point here is that the localization radius should be connected to physical length scales to consider all relevant observations for a gridpoint. However, for a global atmospheric model, the order of the localization radius is 1000 km, the typical size of a low-pressure area, which often contains millions of observations. The only serious solution to this problem is to project the observations onto the ensemble space, but that still does ignore large parts of the observation information.

Of course, this problem does not exist when one does pure parameter estimation, and localization can be a beneficial technique. The first to explore this is Vossepoel and Van Leeuwen, (2007), who used 128 particles to successfully update the order of 10,000 turbulence parameters in a global ocean model.

10.9 Summary

Localization allows us to compute an updated ensemble with realizations outside the space spanned by the initial ensemble. We have discussed several localization methods in this chapter, and the most efficient methods depend on the problem at hand. Localization and inflation are essential tools in high-dimensional data assimilation problems. They are, in practice, used for many other issues than the low-rank prior covariance and spurious correlations between gridpoints far apart. They are also used as tuning parameters to compensate for many problems, such as unknown model errors, approximations in data-assimilation schemes, forward model deficiencies, less well-known observation operators, etc.

Both localization and inflation are, in essence, ad-hoc procedures invented to make the system work. As such, they can introduce so-called unbalanced system states. The classic example is a linear model, where all realizations are solutions, and consequently, linear combinations of the realizations will also be solutions. When applying localization, we break this property and introduce realizations that may not be physically realizable or strong adjustment dynamics to the unbalanced part of the system's state space.

Finally, localization and inflation are two approximate methods to correct related errors in the data assimilation system. Therefore, we must calibrate them to work complementary together.

Methods' Summary

11

The previous chapters illustrate how we could start from Bayes' theorem and apply a sequence of assumptions or approximations that allow us to derive the most popular data-assimilation methods in use today. This chapter attempts to summarize the different techniques and present and compare them in the context of the approximations we made to derive them. We provide a graphical overview that makes it easy to relate different methods and lists the applied approximations.

11.1 Discussion of Methods

The graphical presentation in Fig. 11.1 summarizes all methods, assuming that the underlying dynamical model is nonlinear.

In Chap. 2, we saw that we could split an extended timeline into separate assimilation windows as long as the dynamical model is a Markov process and the measurement errors are uncorrelated between assimilation windows. Thus, we can treat the assimilation problem one window at a time, and the recursive version of Bayes' formula in Eq. (2.23) applies for each assimilation window. We also saw that the parameter-estimation problem is analogous to the assimilation problem for one assimilation window.

As discussed in Sect. 2.3.3, ensemble integrations are a common and probably the only practical means of propagating the state error covariances, or, more generally, the state's pdf, over an assimilation window or from one assimilation window to the next.

Starting from the recursive Bayes' formulation, we can choose two routes. We can solve the Bayesian problem by using a particle representation of the pdf and particle or particle-flow filters to compute the recursive update steps described in Chap. 9. These methods tend to be expensive, and we should only use them when the system

© The Author(s) 2022
G. Evensen et al., *Data Assimilation Fundamentals*,
Springer Textbooks in Earth Sciences, Geography and Environment,
https://doi.org/10.1007/978-3-030-96709-3_11

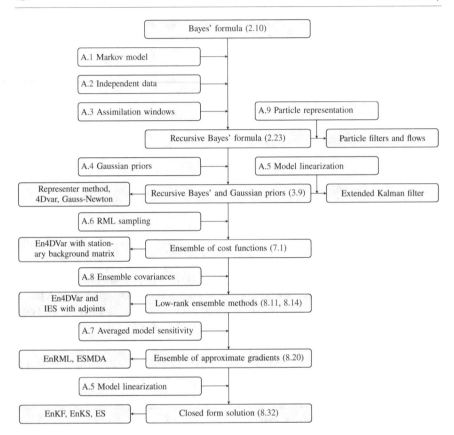

Fig. 11.1 Unified derivation of DA methods. We have summarized the data-assimilation methods and their applied approximations when solving the update over one assimilation window. Other approximations may apply for sequential-in-time data assimilation, e.g., using a stationary background covariance like in 4DVar

is strongly nonlinear. The alternative is to apply the Gaussian priors' assumption in Approx. 4, which allows for deriving most data-assimilation methods currently in use.

The Gaussian-priors' approximation effectively allows us to search for the MAP estimate by minimizing the cost function in Eq. (3.9). Some examples include 4DVar schemes and the representer method, as discussed in Chaps. 4 and 5. Note that these methods only compute the MAP estimate and do not sample the posterior Bayes' distribution. Also, they do not have a direct means for computing and propagating the error statistics to the next assimilation window. Therefore, these methods typically require an additional approximation of a stationary-in-time background-error-covariance matrix for the prior state estimate in each assimilation window.

However, with the correct priors for an assimilation window, if these iterative and adjoint-based methods converge to the cost function's global minimum, they find the MAP solution.

We also saw that, compared to the 4DVar solution, the extended Kalman filter (EKF) applies an additional approximation by linearizing the model and measurement operators to find an explicit update equation. The update only approximates the MAP estimate, but the EKF provides the means for updating and evolving the error statistics in time. The Kalman Filter (KF) solves the data-assimilation problem defined by Eq. (3.8) in the case of a linear model with Gaussian priors. In the weakly nonlinear case with Gaussian priors, the EKF provides an approximate solution due to linearization. However, both KF and EKF require the storage and propagation of the state error-covariance matrix, which becomes overwhelming for real-size geoscience data-assimilation problems, not to mention the severity of the linearization of the error covariance propagation, see Eq. (2.28).

Another alternative route is to follow the Gaussian-priors assumption with the Randomized Maximum Likelihood sampling approach described in Chap 7, providing an approximate sampling of the posterior pdf. The RML sampling requires minimizing an ensemble of cost functions with different prior state vectors and perturbed measurements. RML sampling turns out to be exact in the Gauss-linear case. At the same time, with significant nonlinearities, we cannot expect that it will work satisfactorily, and we might need to use the particle or particle-flow filters.

The RML sampling allows us to derive several ensemble-based assimilation methods. One alternative would be to use the 4DVar schemes to minimize each cost function to provide an RML ensemble of solutions. En4DVar uses adjoints and solves each cost function realization exactly, at least as long as there are no local minima. Hence, En4DVar solves the RML sampling problem within the approximations made on the background matrix. This approach is close to the procedure used by some operational En4DVar systems. But the method still largely ignores updating and evolving the error statistics.

Using a sufficiently large ensemble makes it possible to compute posterior error statistics and propagate error statistics from one time window to the next. We can then use the forecast ensemble to compute the prediction error covariance and use it instead of the stationary background matrix. Thus, since we are not able to calculate the full exact covariances, Approx. 8, where we compute the covariances from the ensemble, allows us to design methods with consistent error statistics evolving in time. Both En4DVar, using either the strong-constraint algorithms or the weak-constraint representer method and other adjoint-based Gauss–Newton methods, would work in this configuration. As the standard En4DVar uses a stationary background matrix, a complete ensemble-based En4DVar should outperform the traditional version as long as we use a sufficiently large ensemble.

If an adjoint model is available, we would not need to introduce additional approximations. However, in many cases, the adjoint model does not exist, and with commercial software, we often do not have access to the model code, so we cannot implement the adjoint model. Also, in this case, there is an alternative that uses the averaged model sensitivity from Approx. 7. We replace the individual adjoints for each realization with an ensemble-averaged model sensitivity. This best linear fit of the model sensitivity is the same for all the realizations. Thus for nonlinear systems, we introduce an approximation by changing each realization's gradient slightly. Using an averaged model sensitivity leads to modern and efficient methods like EnRML and ESMDA, discussed in Chap. 8. For instance, the petroleum industry uses these methods operationally to history match reservoir models.

EnRML and ESMDA are iterative methods and require multiple integrations of the ensemble over the assimilation window. A computationally more efficient approach is to use the ensemble smoother (ES) for the assimilation window. The ES introduces a linearization of the model in the expression for the gradient, which allows for deriving a closed-form solution that we can compute without iterating. This equation is only valid for minor updates or modest nonlinearity. This property is precisely the basis for ESMDA. By calculating many minor updates using the ES equation instead of a single large one, ESMDA reduces the impact of the linearization.

In the "trivial" case with linear models and measurement operators, Fig. 11.1 would simplify the algorithms significantly. The use of Gaussian priors ensures that the distribution at all future times is also Gaussian. Furthermore, the RML sampling is precisely sampling the posterior pdf and does not introduce any approximation. There will be sampling errors, of course, since we use a low-rank ensemble approximation. The averaged model sensitivity is exact in this case, and there is no need for any linearizations or iterations.

11.2 So Which Method to Use?

It is impossible to provide specific advice on which method to use for a particular data-assimilation problem in a book like this one. There are just too many different problems out there. But, perhaps, more importantly, even the experts can disagree on the best method based on their favorite techniques and experience. However, we can offer some general advice. At the same time, the user has to keep in mind that fine-tuning a data-assimilation method remains an art, as it is true with any valuable thing in life.

The choice of method depends on the data assimilation application and its purpose. We will come back to this question in the final Chap. 23. But first, it is essential to evaluate the nature of the system, as this will determine the efficacy of the various data-assimilation methods. Thus, in this book's Part II, we explain several example applications to demonstrate how different assimilation methods work with various dynamic models and assimilation problems. We will present examples that illustrate smoother versus sequential estimation, state versus parameter estimation, weak-constraint versus strong-constraint solutions, and linear versus nonlinear or even highly nonlinear problems.

Part II
Examples and Applications

The second part of this book presents several simple examples and applications to demonstrate the properties of different data-assimilation methods. The purpose is to add context to the theoretical discussion in Part I. The examples range from simple examples with linear dynamics to highly nonlinear cases. We will also present two real-world applications to illustrate the methods' potential: a high-dimensional parameter-estimation problem solved in the petroleum industry. and an application of data assimilation for parameter- and control-variable estimation with an epidemic model for COVID-19.

A Kalman Filter with the Roessler Model

12

In this chapter, we discuss an application of the Kalman filter on simple systems to study its behavior in linear and nonlinear situations. We will use the Roessler model, which we can configure as both a linear and nonlinear system. We start with the linear system, for which the Kalman filter provides an exact and optimal solution, and then study the extended Kalman filter's performance in the nonlinear system.

12.1 Roessler Model System

The governing equations defining the Roessler model system for the three variables x, y, and z are

$$\frac{\partial x}{\partial t} = -y - z, \tag{12.1}$$

$$\frac{\partial y}{\partial t} = x + ay, \tag{12.2}$$

$$\frac{\partial z}{\partial t} = b - cz + \beta xz, \tag{12.3}$$

in which β is a parameter that determines the system's nonlinearity. If $\beta = 1$, the system has stable, periodic, and chaotic solutions, depending on the values of the three other parameters. Its bifurcation diagrams are similar to that of the logistic map. Choosing $\beta = 0$ results in a linear system with either growing, decaying, or periodic solutions.

© The Author(s) 2022
G. Evensen et al., *Data Assimilation Fundamentals*,
Springer Textbooks in Earth Sciences, Geography and Environment,
https://doi.org/10.1007/978-3-030-96709-3_12

12.2   Kalman Filter with the Roessler System

We first look at the behavior of the Kalman filter from Chap. 6 applied to a linear version of the Roessler system resulting by setting $\beta = 0$. The system matrix $\mathbf{M}$ then becomes

$$\mathbf{M} = \begin{pmatrix} 0 & -1 & -1 \\ 1 & a & 0 \\ 0 & 0 & -c \end{pmatrix}, \tag{12.4}$$

and we have chosen $b = 0$ to simplify the analysis.

We can find the eigenvalues of this linear system via the characteristic equation $\det(\mathbf{M} - \lambda \mathbf{I}) = 0$, leading to

$$\lambda = -c \quad \vee \quad \lambda = \frac{a \pm \sqrt{a^2 - 4}}{2}. \tag{12.5}$$

We will always define a $c > 0$, so the first eigenvalue corresponds to a decaying mode. Positive values of a lead to ever-growing modes that can be spiraling for $a < 2$ and growing exponentially for $a \geq 2$. Negative values for a lead to spiraling decaying modes. In the following, we will use $a = 0$, so a purely oscillatory mode.

We generate the reference solution by solving the system using a fourth-order Runge–Kutta (RK4) time-stepping scheme with a time step of $\Delta t = 0.1$. For the data-assimilation experiment, we use the same time step but employ an Euler Scheme. The Euler scheme is less accurate and will give rise to a model error at every time step.

Figure 12.1 gives the evolution for the 3-dimensional system over 200 time steps starting from initial condition $(6, 0, 0)$, for the RK4 and the Euler scheme over 500 time steps. The parameter setting is $(a, b, c) = (0.0, 0.5, 0.1)$. The true solution converges towards a periodic orbit, while the Euler solution spins out of control. Thus, we need to use data assimilation to keep the system on track!

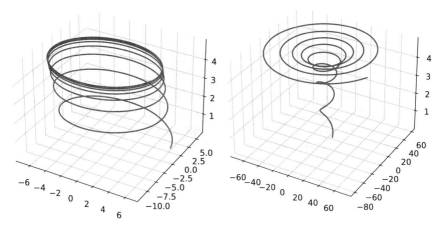

Fig. 12.1 Evolution of the linear system without data assimilation using an accurate RK4 scheme (left) and a less precise Euler scheme (right) with the same time step

The Kalman filter needs evolution equations for both the mean and the covariance. The evolution equation for the mean is simply the forward model using the Euler scheme, i.e.,

$$\mathbf{x}_{k+1} = \mathbf{x}_k + \Delta t \mathbf{M} \mathbf{x}_k = (\mathbf{I} + \Delta t \mathbf{M}) \mathbf{x}_k, \tag{12.6}$$

which corresponds to the model defined in Eq. (6.34). We can derive the evolution equation for the covariance Eq. (6.35) by first subtracting the true solution at time $k + 1$ from Eq. (12.6), leading to

$$\begin{aligned} \mathbf{x}_{k+1} - \mathbf{x}_k^t &= (\mathbf{I} + \Delta t \mathbf{M}) \mathbf{x}_k - \mathbf{x}_{k+1}^t \\ &= (\mathbf{I} + \Delta t \mathbf{M})(\mathbf{x}_k - \mathbf{x}_k^t) + \mathbf{q}_k \end{aligned}, \tag{12.7}$$

in which $\mathbf{q}_k$ denotes the difference between using the RK4 or the Euler scheme. If we now multiply the equation from the right with $(\mathbf{x}_{k+1} - \mathbf{x}_k^t)^T$, expand the right-hand side, and take the expectation, we find

$$\mathbf{C}_{xx,k+1} = (\mathbf{I} + \Delta t \mathbf{M}) \mathbf{C}_{xx,k} (\mathbf{I} + \Delta t \mathbf{M})^T + \mathbf{C}_{qq,k}, \tag{12.8}$$

where we assumed the model errors to be independent of the errors in the model state. Finally, keeping only terms up to Δt, consistent with the Euler scheme, we arrive at

$$\mathbf{C}_{xx,k+1} = \mathbf{C}_{xx,k} + \Delta t (\mathbf{M} \mathbf{C}_{xx,k} + \mathbf{C}_{xx,k} \mathbf{M}^T) + \mathbf{C}_{qq,k}, \tag{12.9}$$

which is the evolution equation for the covariance we will use in our Kalman filter.

In our first experiment, we use observations of all three variables every 5th time step, with observation error standard deviation $\sigma_{obs} = 0.1$, over 1000 timesteps. Furthermore, the model error covariance matrix $\mathbf{C}_{qq}$ is diagonal with diagonal elements equal to 0.01. We give the results of the time evolution over 1000 time steps in Fig. 12.2. We see that the Kalman filter solution wiggles around the true solution and is stable. Figure 12.3 shows the evolution of the root-mean-square error in black

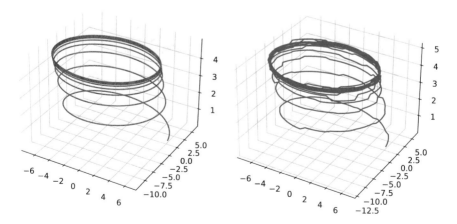

Fig. 12.2 True solution (left) and Kalman filter solution (right) with all variables observed every 5 time steps

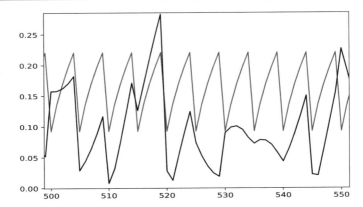

Fig. 12.3 Time evolution of the RMSE (black) and estimated error standard deviation of 1st variable (blue) with all variables observed every 5th time step

and the Kalman filter uncertainty of the first variable in blue. The blue line shows the characteristic behavior of Kalman filter estimated errors, with growth between observation errors and a strong reduction at assimilation time. The black line looks rather chaotic, which is not unexpected because it is a random variable. We see that the error estimation is successful as the blue and black lines have similar means. It is also interesting to note that the minimum estimated error is slightly below the observation error, as expected when the state error is larger than the observation error.

The following experiment has the same settings as before, but we only observe the first variable every five time steps. The solution looks quite similar in Fig. 12.4, which shows that information from the observed variable also affects the other two variables. Two processes can be responsible for this transfer. The first process is the

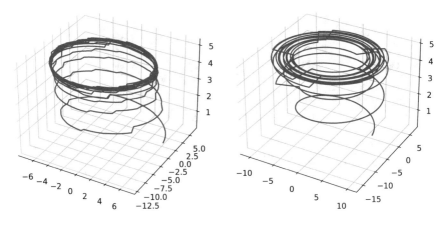

Fig. 12.4 Kalman filters solutions for only 1st variable observed every 5th time step (left) and all variables observed every 100th time-step (right)

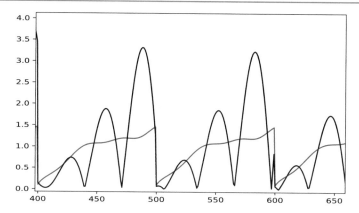

Fig. 12.5 Time evolution of the RMSE (black) and estimated error standard deviation of 2nd variable (blue)

update of unobserved variables via their covariance with the observed variables. Just before the last assimilation step, we found the relative magnitude of that correlation as 0.2 and 0.1 with the 2nd and the 3rd variable, respectively. These correlations are low, and the variance in unobserved and observed variables differs by about a factor of 10. Hence, the second process, i.e., the connection via the model dynamics, is also essential. The periodic orbit is a stable solution of the system, and forcing one of the variables to follow the true evolution draws in the others to do the same. The root-mean-square error confirms the importance of this effect in the 2nd and 3rd variables, which is about a factor $\sqrt{10}$ lower than the estimated error from the covariance matrix.

Figure 12.4 also shows the Kalman filter solution for observations every 100 time steps, approximately once every two revolutions, again observing all variables. While the estimate is not perfect, it remains stable and keeps tracking the reference evolution, although at some distance. We confirm this result by the parallel development of the true and the estimated error in the 2nd variable in Fig. 12.5. Note that the errors grow considerably between observation times but return to the observation error value at assimilation times.

These experiments demonstrate the strength of the Kalman filter in linear systems. Our next set of experiments concerns the extended Kalman filter applied to the nonlinear Roessler system.

12.3   Extended Kalman Filter with the Roessler System

In this section, we study the application of the Kalman filter, or rather the Extended Kalman filter, to the full nonlinear Roessler model, where we now set $\beta = 1$. Figure 12.6 shows the evolution of this system using a Runge–Kutta 4 scheme for 2000

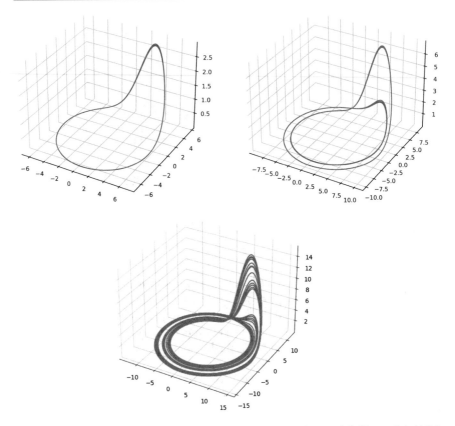

Fig. 12.6 Examples of the time evolution of the Roessler system for $c = 4$ (left), $c = 6$ (middle), and $c = 9$ (right), showing, respectively periodic behavior with periods one and two, and chaotic behavior

time steps with time step $\Delta t = 0.01$, and initial condition $(6, 0, 0)$. We have chosen $a = b = 0.1$, and c varies from 4 to 6 to 9, corresponding to a one-period solution, a two-period solution, and a chaotic solution. We can recognize the remnants of the periodic orbit of the linear system in the x-y plane and the burst into the z plane resulting from the nonlinearity. Also, note the remnants of the one-period solution in the two-period solution, but with enhanced amplitude. Likewise, we see the remnants of the two-period solution in the chaotic case.

We perform the data-assimilation runs using the following parameters settings for the full Roessler system: $(a, b, c) = (0.1, 0.1, 9)$, which brings us into the chaotic regime depicted in the right plot in Fig. 12.6. The time step is $\Delta t = 0.01$, and we choose a diagonal model error covariance matrix $\mathbf{C}_{qq}$ with 10^{-4} on the diagonal. The value of this model error depends on the system setup and we find this value by trial and error as there is no general guideline for such nonlinear systems as the Roessler model. We observe the x variable every 100 time steps, but not the other variables. The observation error standard deviation is 0.1.

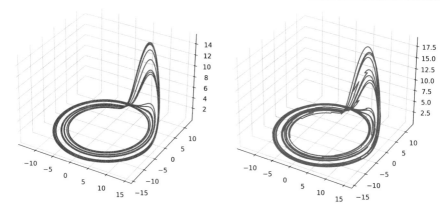

Fig. 12.7 True evolution (left) and mean of the EKF solution (right). Note the difference in scales

We provide an overview of the true solution and the mean of the EKF in Fig. 12.7. The EKF solution does perform well in the x-y plane but tends to overshoot and be less accurate in general in the z variable. This result is not surprising as the z direction is where the nonlinearity in the system has maximal effect via the xz term in Eq. (12.3). The jumps in the solution at assimilation time are visible and not restricted to the x variable only.

Figure 12.8 provides a more detailed view of the EKF performance. The observed variable x and the unobserved y variable remain very close to the true solution,

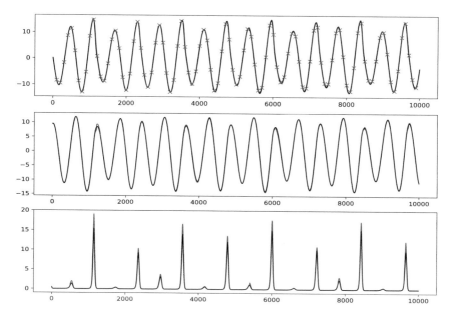

Fig. 12.8 Evolution of x (top), y (middle), and z (botom). The truth (black), observations (red crosses), and the EKF mean (blue) are displayed. Note the large mismatch in the z variable

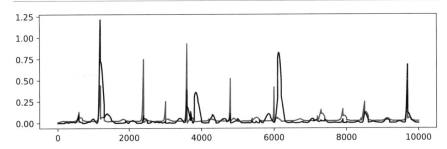

Fig. 12.9 Evolution of the RMSE of x (black) and its estimated standard deviation (blue). Note that while the RMSE remains bounded the estimated error grows exponentially

but the z variable has more significant deviations from the truth, typically slightly overshooting.

One advantage of the EKF is that it does provide an error estimate. We present the root-mean-square error (RMSE) of the second variable y and its estimated standard deviation in Fig. 12.9. The RMSE remains bounded, and the estimated error is of the same order of magnitude, but in general, not accurate at higher values. This result points to a general weakness of the EKF for strongly nonlinear systems. The linearization of the evolution equation for the error covariance can lead to inaccurate and unbounded error growth. This result supports the findings of Evensen (1992), who experienced unbounded error growth when applying the EKF with a nonlinear QG model. The error estimates for the other variables behave similarly.

Linear EnKF Update

13

The Kalman filter or its ensemble version, the ensemble Kalman filter, is optimal for a linear model and -measurement operator. This chapter will comprehensively discuss the EnKF analysis scheme and its properties, focusing on an ensemble-subspace computation of the inverse. We demonstrate the importance of taking into account correlations in the model errors. Furthermore, we study the efficient ensemble-subspace inversion method that allows computing the analysis update at a linear cost in both the number of measurements and the state dimension. We also show how to reduce sampling errors by increasing the number of measurement-perturbations realizations used to represent the measurement-error-covariance matrix.

13.1 EnKF Update Example

We will use an example from Evensen (2021) in the following discussion. The purpose of this example is to illustrate the properties of the update scheme of the ensemble Kalman filter (EnKF) as described in Chap. 8, when using the measurement perturbations to represent the measurement error covariance matrix. In addition, this example verifies the robustness of the projection of the measurement error covariance matrix onto the ensemble of predicted measurements. Finally, the update is identical to the EnRML algorithm's solution in the linear case, and as such, the results are representative of the EnRML smoother update.

The test example uses a one-dimensional periodic domain with 1024 gridpoints and $\Delta x = 1$. In this domain, we simulate a smooth pseudo-random function with mean $\mu = 4$, variance $\sigma^2 = 1$, and decorrelation length $r_d = 40$, representing the unknown truth,

$$\mathbf{z}_{\text{true}} \sim N\left(\mu = 4, \sigma^2 = 1, r_d = 40\right). \tag{13.1}$$

The constant $\mu = 4$, is just added for plotting purposes.

© The Author(s) 2022
G. Evensen et al., *Data Assimilation Fundamentals*,
Springer Textbooks in Earth Sciences, Geography and Environment,
https://doi.org/10.1007/978-3-030-96709-3_13

The first guess solution is generated by simulating another realization $\mathbf{z} \sim \mathcal{N}(0, 1, 40)$ and adding it to the truth, i.e.,

$$\mathbf{z}_{\text{fg}} = \frac{\mathbf{z} + \mathbf{z}_{\text{true}} - 4}{\sqrt{2}} + 4. \tag{13.2}$$

The factor $\sqrt{2}$ ensures that the variance of $\mathbf{z}_{\text{fg}}$ is equal to one.

The initial ensemble is created by adding random realizations $\mathbf{z}_j \sim \mathcal{N}(0, 1, 40)$ to the first guess $\mathbf{z}_{\text{fg}}$,

$$\mathbf{z}_j^{\text{f}} = \mathbf{z}_{\text{fg}} + \mathbf{z}_j. \tag{13.3}$$

The measurements are distributed uniformly over the domain and sampled from a perturbed true solution according to

$$\mathbf{d}_j = \mathbf{H}(\mathbf{z}_{\text{true}} + \mathbf{z}_j), \tag{13.4}$$

with either uncorrelated, $\mathbf{z}_j \sim \mathcal{N}(0, 0.25, 0)$, or Gaussian, $\mathbf{z}_j \sim \mathcal{N}(0, 0.25, 40)$, perturbations. Here, $\mathbf{H}$ is the linear measurement operator that extracts the measurements from the functions $\mathbf{z}_{\text{true}} + \mathbf{z}_j$.

The following experiments use the same reference truth, measurements, initial ensemble, and random seed. However, when it is essential to eliminate sampling errors, we use extended ensemble sizes. Thus, although there are seed dependencies in the obtained solutions, the different methods should produce the same 'nswer. Furthermore, in most experiments, we can attribute differences in the results to the methods used. Thus, our approach is different from running multiple data-assimilation experiments with varying seeds.

13.2   Solution Methods

In the following, we study two prominent cases, one with a diagonal measurement error-covariance matrix and another with correlated errors, $\mathbf{e}_i$, simulated from Eq. (13.4) and with $r_d = 40$. For the two cases of uncorrelated and correlated measurement errors, the EnKF computes the analysis using either an exactly specified measurement error-covariance matrix $\mathbf{C}_{dd}$ or representing the measurement error covariance by the perturbations in $\mathbf{E}$. Since $\mathbf{H}$ is a linear operator, we follow the EnKF update Eq. (7.3), rather than Eq. (7.12) because for this linear example, we do not need to consider the modification in Eq. (7.11).

The case with a full-rank measurement error-covariance matrix solves

$$\mathbf{X}^{\text{a}} = \mathbf{X}^{\text{f}} + \mathbf{A}\mathbf{S}^{\text{T}}\left(\mathbf{S}\mathbf{S}^{\text{T}} + \mathbf{C}_{dd}\right)^{-1}(\mathbf{D} - \mathbf{H}\mathbf{X}), \tag{13.5}$$

where the ensemble perturbation matrix is $\mathbf{A} = \mathbf{X}\mathbf{\Pi}$. The matrix $\mathbf{C} = \mathbf{S}\mathbf{S}^{\text{T}} + \mathbf{C}_{dd}$ is formed and then inverted by computing an eigenvalue decomposition $\mathbf{C} = \mathbf{Q}\mathbf{\Lambda}\mathbf{Q}^{\text{T}}$. The inverse is just $\mathbf{C} = \mathbf{Q}\mathbf{\Lambda}^+\mathbf{Q}^{\text{T}}$ where the use of a pseudo inverse is needed in case the matrix $\mathbf{C}$ is poorly conditioned.

When using an ensemble representation for the measurement error-covariance matrix $\mathbf{C}_{dd} = \mathbf{E}\mathbf{E}^{\mathrm{T}}$, we can solve for the update from

$$\mathbf{X}^{\mathrm{a}} = \mathbf{X}^{\mathrm{f}} + \mathbf{A}\mathbf{S}^{\mathrm{T}}\left(\mathbf{S}\mathbf{S}^{\mathrm{T}} + \mathbf{E}\mathbf{E}^{\mathrm{T}}\right)^{-1}(\mathbf{D} - \mathbf{H}\mathbf{X}). \tag{13.6}$$

In the examples below, the line labels used in the figures indicate the scheme used to compute the matrix inversion. The line label Cdd denotes using the standard EnKF analysis equation with a full rank measurement error-covariance matrix $\mathbf{C}_{dd}$, as explained above. The curves with line label EE correspond to the EnKF update when the samples in $\mathbf{E}$ replace the "exact" analytic measurement error covariance matrix $\mathbf{C}_{dd}$, and we use the ensemble subspace scheme.

Using $\mathbf{E}$ to represent the measurement error-covariance matrix introduces additional sampling errors. However, we will see below how to reduce these sampling errors to a negligible level with a simple algorithm modification. I.e., one uses a larger number of realizations in $\mathbf{E}$ to represent $\mathbf{C}_{dd}$ better. The code used is the test case from https://github.com/geirev/EnKF_analysis.git.

13.3   Example 1 (Large Ensemble Size)

The first example uses 50 measurements and a large ensemble size of 2000 to reduce sampling errors. Figure 13.1 shows the results for the two cases with either diagonal or correlated measurement errors.

The upper-left plot shows the EnKF estimates for the case with uncorrelated measurement errors for each grid cell numbered with indexes 1–1024. The two schemes, represented by the lines labeled Cdd and EE, give similar results in this case. The upper-right plot shows the prior and posterior error variances for the two updates, and again they are nearly identical. Finally, the lower-left plot presents the EnKF estimates for the case with correlated measurement errors. In this case, we also see that the results using the exact and approximate schemes (Cdd and EE) are nearly identical.

An apparent difference between the two cases is that, with uncorrelated errors, the measurements are scattered randomly about the correct solution. In contrast, with correlated measurement errors, successive measurements will have similar error values, and they follow a smooth curve. The nonzero measurement correlations' role is to reduce the strength of the update, and the result is an update with a more substantial variance. By taking the measurement error correlations into account, we inform EnKF that neighboring measurements make the same error and reduce their accumulated impact.

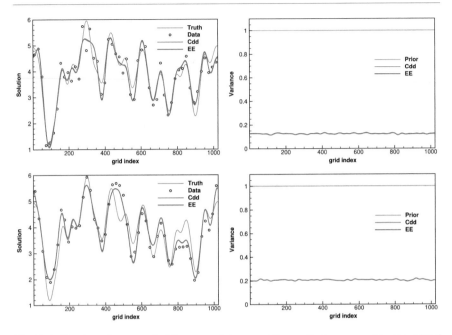

Fig. 13.1 Simple update example: The upper plots present the results for a case with uncorrelated measurement errors, while the lower plots give the results when using measurements with correlated errors and decorrelation length $r_d = 40$. The left plots show the results for the posterior ensemble means, while the right plots provide the associated error estimates. The line labels Cdd, EE, denote different numerical implementations of the inversion scheme used, as is explained in the text. The ensemble size is 2000

13.4 Example 2 (Ensemble Size of 100)

We now repeat the previous experiment from Example 1 using a more common ensemble size of 100. The purpose is to illustrate the impact of sampling errors when using the measurement error perturbations in **E** to represent $\mathbf{C}_{dd}$. Figure 13.2 shows that with 100 realizations, the additional sampling errors introduced by scheme EE lead to a slight deviation between the two estimates. More problematic is the underestimation of the ensemble variance. In a sequential data-assimilation context, this underestimation would have to be compensated for, e.g., by using inflation, to avoid possible filter divergence. In the following example, we will learn how to reduce these sampling errors to a negligible level.

13.5 Example 3 (Augmenting the Measurement Perturbations)

The benefit of using Eq. (13.6) over Eq. (13.5) is the reduced computational cost, but also the fact that it is easier to sample perturbations with accurate statistics than constructing a full-rank measurement error covariance matrix. An approach for re-

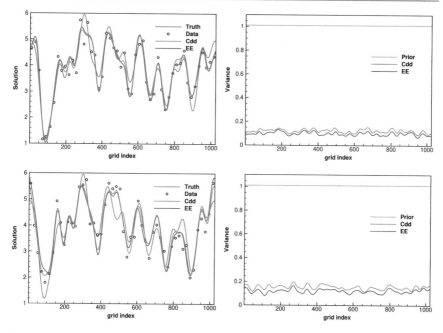

Fig. 13.2 Simple update example: Same as Fig. 13.1 but using an ensemble size of 100 realizations

ducing the sampling errors in scheme EE is to augment columns of new realizations of measurement perturbations to **E**. This modification only slightly increases the computational cost of the algorithm when computing $\mathbf{\Sigma}^{+}\mathbf{U}^{T}\mathbf{E}$ in Eq. (8.48) and is a simple modification of the code. Figure 13.3 shows the results using 100 realizations and correlated measurement errors, and when using 1000 samples in **E**. The augmentation of additional columns to **E** in Exp. 3 significantly reduces the errors in the estimated means and variances for the two cases with correlated and uncorrelated measurement errors compared with the results from Exp. 2. It is clear that the two schemes Cdd and EE, solving Eqs. (13.5) and (13.6) respectively, give almost identical results. In this case, the measurement error perturbations' projection onto the ensemble subspace does not significantly impact the results. Thus, the sampling errors introduced by using **E** to represent $\mathbf{C}_{dd}$ can be made negligible by increasing the sample size in **E** to only a minor additional cost.

Evensen (2021) found that the algorithm works well with different measurement error decorrelation lengths. When the measurement perturbations include small scales not represented by the predicted measurements' ensemble, the projection onto the ensemble anomalies in **S** introduces an approximation. The truncation of small scales in the measurement errors leads to a slight underestimation of the measurement error variance.

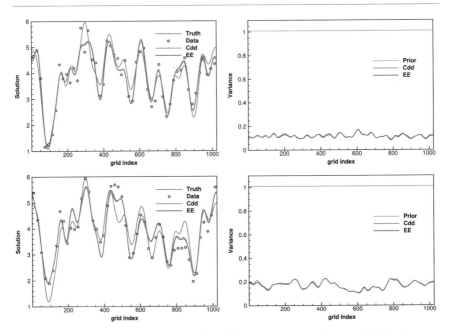

Fig. 13.3 Simple update example: Same as Fig. 13.2 but using an ensemble size of 1000 realizations to represent **E**

13.6   Example 4 (Large Number of Measurements)

Figure 13.4 shows results from the final example where the number of measurements increased from 50 to 200, i.e., twice the ensemble size. In this case, we apply a truncation at 99% of the variance when computing the inversion, retaining 29 singular values when computing the singular value decomposition of **S**. Again, the results obtained are very similar using the two algorithms. It is also interesting to see how

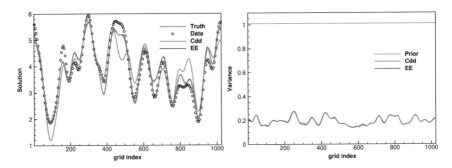

Fig. 13.4 Simple update example: Same as Fig. 13.3 but for a case with 200 measurements, which is twice the ensemble size, and with measurement-error correlations $r_d = 40$

the measurements' impact reduces at the grid cells with indices 400–500. Note that there is no indication of the so-called "ensemble degeneracy," and the analysis ensemble retains a significant variance. The posterior variance using 200 measurements is similar to the one obtained using only 50 observations. This result indicates that including additional dependent measurements does not introduce much new information in this example.

EnKF for an Advection Equation

14

This chapter discusses a straightforward application of the EnKF with a linear advection equation. The example illustrates the smooth spatial update that the EnKF provides and how information propagates with the flow. Furthermore, we will see how the EnKF provides consistent error statistics.

14.1 Experiment Description

We now present an extension of the linear update example from Chap. 13 using the EnKF described in Chap. 8, including a recursion in time. We consider a linear advection model on a periodic one-dimensional domain where a wave propagates from left to right. As in the example of Chap. 13, we define a non-dimensional grid with $\Delta x = 1$ and 1024 grid cells. Also, we define a non-dimensional time step $\Delta t = 1$ and constant advection velocity $u = 1$.

Thus, we now consider the data assimilation on a rectangular space-time domain where the model dynamics propagates the information forward in time from one assimilation update to the next, subject to stochastic model errors sampled from $\mathcal{N}(0, 0.0009, 20)$. With a linear dynamical model, any of the error-propagation equations will give the same result (with the ensemble approach requiring an infinite ensemble size). We have used an EnKF implementation with 5000 realizations, which is large enough to make the EnKF and KF solutions practically indistinguishable. After each update we integrate the updated ensemble forward, subject to stochastic noise until the next update step.

We can relate this problem to a formulation with multiple assimilation windows defined from a measurement time up to and including the next measurement time. We also consider the filtering problem described in Sect. 2.4.2, where we update

© The Author(s) 2022
G. Evensen et al., *Data Assimilation Fundamentals*,
Springer Textbooks in Earth Sciences, Geography and Environment,
https://doi.org/10.1007/978-3-030-96709-3_14

the solution at the end of the assimilation window. This update serves as the initial condition for further integration through the next assimilation window.

As in the previous section, we simulate a true pseudo-random wave sampled from $\mathbf{z}_{\text{true}} \sim \mathcal{N}(\mu = 4, \sigma^2 = 1, r_d = 20)$. We also generate the initial ensemble using the procedure from Chap. 13, starting from a first guess field $\mathbf{z}_{\text{fg}}$ computed using Eq. (13.2) with $\mathbf{z} \sim \mathcal{N}(0, \sigma^2 = 1, r_d = 20)$. Then we add the ensemble of perturbations $\mathbf{z}_j \sim \mathcal{N}(0, 1, 20)$ to the first-guess to create the initial ensemble from Eq. (13.3).

To generate the measurements, we integrate the model forward using the exact dynamical model starting from the true solution at $t = 0$. We measure the true solution at 10 equidistributed spatial locations at every $\Delta t = 5$, and we add uncorrelated noise to the measurements sampled from $\epsilon \sim \mathcal{N}(0, \sigma^2 = 0.04)$.

14.2   Assimilation Experiment

Figure 14.1 presents three snapshots at $t = 10$, $t = 50$, and $t = 100$, hence after 2, 10, and 20 updates. The thick blue line is the ensemble mean, with the light blue shading representing plus and minus two standard deviations of the ensemble variance. We also show the yellow circles with their two standard-deviation error bars representing the measurements. The green line at the bottom of the plots is the estimated ensemble standard deviation. We have also included a thin dashed line, which is the true solution to recover through the data assimilation.

From the plots we notice how the assimilation of measurements pulls the solution closer to the measured value at the measurement locations. Simultaneously, the ensemble spread and error variance reduce. From the green line we also notice how the information from the measurements propagates downstream. As the model is also subject to a stochastic system noise or model error, there is a linear increase in error variance downstream of each measurement. The final plot is a quasi-stationary solution where the variance increase resulting from the introduced stochastic forcing balances the variance reduction from the assimilation updates. Note that the updated ensemble's standard deviation at each observation point is similar to the observation error standard deviation of 0.2, as expected from Kalman filter theory. The code used is available from https://github.com/geirev/EnKF_advection.git.

In the following chapters, we will introduce nonlinearities to the assimilation problem, which makes the assimilation process and the interpretation of the results more complicated.

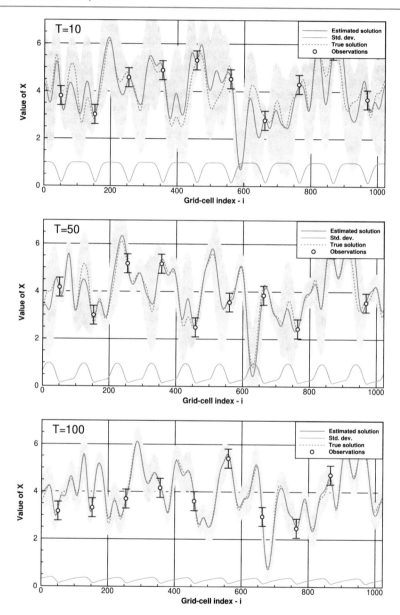

Fig. 14.1 The plots show a time sequence of estimates from using EnKF with a linear advection equation. The thick blue line is the ensemble mean, while the blue shadow illustrates the ensemble spread by showing plus-minus two ensemble standard deviations. The reference solution is the dashed blue line, and the observations are the yellow dots, including an error bar indicating plus-minus two standard deviations. The green line is the estimated ensemble standard deviation. This example is further illustrated in the animation available from https://github.com/geirev/EnKF_advection/blob/master/doc/animation.gif

EnKF with the Lorenz Equations

15

The chaotic Lorenz'63 model is a much-used testbed used to examine the capabilities of data-assimilation methods to handle nonlinear, unstable, and chaotic dynamics. This chapter will repeat some experiments that demonstrate the strengths of ensemble methods for highly nonlinear dynamics. We mainly focus on applying three different ensemble methods, the ensemble smoother (ES), ensemble Kalman smoother (EnKS), and ensemble Kalman filter (EnKF).

15.1   The Lorenz'63 Model

We will now repeat an example from Evensen (1997) and Evensen & Van Leeuwen (2000) with the chaotic Lorenz (1963) model to demonstrate the properties of three different ensemble methods, the ensemble smoother (ES), ensemble Kalman smoother (EnKS), and ensemble Kalman filter (EnKF), when applied to highly nonlinear dynamics. The famous Lorenz (1963) model is a coupled system of three nonlinear ordinary differential equations

$$\frac{\partial x}{\partial t} = \sigma(y - x), \tag{15.1}$$

$$\frac{\partial y}{\partial t} = \rho x - y - xz, \tag{15.2}$$

$$\frac{\partial z}{\partial t} = xy - \beta z. \tag{15.3}$$

Here $x(t)$, $y(t)$, and $z(t)$ are the dependent variables, and we use the common parameter values $\sigma = 10$, $\rho = 28$, and $\beta = 8/3$. The initial conditions for the reference case are $(x(0), y(0), z(0)) = (1.508870, -1.531271, 25.46091)$ and the time interval is $t \in [0, 40]$.

© The Author(s) 2022
G. Evensen et al., *Data Assimilation Fundamentals*,
Springer Textbooks in Earth Sciences, Geography and Environment,
https://doi.org/10.1007/978-3-030-96709-3_15

We generate the observations and initial conditions by adding normally distributed white noise with zero mean and variance equal to 2.0 to the reference solution. We also measure all the variables x, y, and z at regular time intervals $\Delta t = 0.5$. This value is half the typical revolution time in each of the wings of the system. The ensemble size is $N = 2000$. The current setup corresponds to Experiment B from (Evensen, 1997). In the upper plots in Figs. 15.1, 15.2 and 15.3, the red line denotes the estimate and the blue line is the reference solution. In the lower plots the red line is the standard deviation estimated from ensemble statistics, while the blue line is the absolute value of the true residuals with respect to the reference solution.

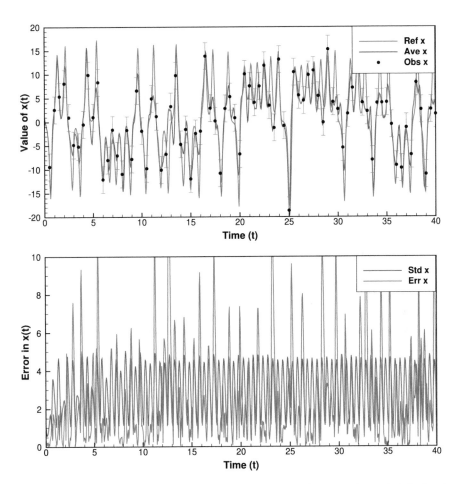

Fig. 15.1 Ensemble Smoother (ES). The upper plot shows the inverse estimate *(red line)* and reference solution *(blue line)* for x. The black points are the measurements with error bars showing the plus and minus two standard deviation range. The lower plot shows the corresponding estimated standard deviations from the ensemble *(red line)* as well as the actual posterior errors, *(blue line)*, computed as the absolute value of the difference between the reference solution and the ensemble mean estimate

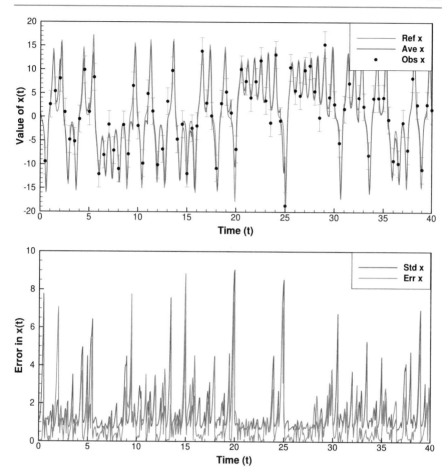

Fig. 15.2 Ensemble Kalman filter. See Fig. 15.1 for description

15.2 Ensemble Smoother Solution

The ensemble smoother (ES) does not split the assimilation interval into multiple assimilation windows but rather attempts to tackle the smoother formulation from Sect. 2.4.1. The key to the ES method is using an unconstrained ensemble integration over the whole assimilation interval that represents the prior error statistics. After that, all measurements are processed in one go to produce the posterior estimate. Thus, the ES attempts to solve Bayes' theorem in Eq. (2.10). The ES uses Approx. 4, by assuming a Gaussian prior, but also Approx. 8 by using a finite but large ensemble size. With this approach, we do not need to make any of the Approxs. 1 (Markov model), 2 (uncorrelated measurement errors in time), or 3 (filtering approximation). Furthermore, for a linear measurement operator, as in the current example, the linearization, the sampling, and the linear regression approximations 5, 6, and 7 do

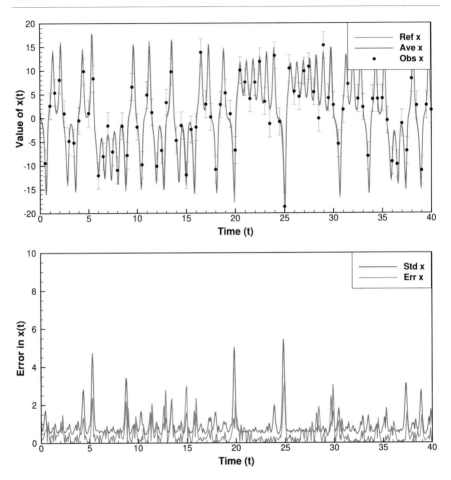

Fig. 15.3 Ensemble Kalman smoother. Ensemble Kalman filter. See Fig. 15.1 for description

not apply. Interestingly, for a linear model with Gaussian statistics, ES provides the optimal solution. However, for a highly nonlinear model like the Lorenz equations, the Gaussian assumption for the prior is severe.

Figure 15.1 presents the ES solution and estimated error variance for the x-component. ES performs rather poorly for this example. However, even if the fit to the reference trajectory is poor, the ES solution captures most of the transitions. The main problem is to estimate of the amplitudes in the reference solution. We attribute the cause for the weak performance to non-Gaussian contributions in the prior distribution for the model evolution, as can be expected in such a strongly nonlinear case.

The error estimates evaluated from the posterior ensemble are not large enough at the peaks where the smoother performs poorly. The underestimated errors result from neglecting the non-Gaussian contribution from the probability distribution for the model evolution. Otherwise, the error estimate looks reasonable with minima at the measurement locations and maxima between the measurements.

15.3 Ensemble Kalman Filter Solution

As for the advection example in Chap. 14 we now apply the Approxs. 1 (Markov model), 2 (uncorrelated measurement errors in time), and 3 (assimilation time-window approximation). Furthermore, in addition to the Approx. 4 (Gaussian priors) and 8 (the ensemble representation), we compute the filter solution by only updating the state at the end of the time window. As when using ES, the linearization, the sampling, and the linear regression approximations 5, 6, and 7 do not apply. As for the advection case in Chap. 14, we update the solution at the end of each time window, using the standard EnKF update equation in a filtering configuration.

EnKF does an excellent job tracking the reference solution as shown in Fig. 15.2 and captures all transitions. For example, at $t = 9, 11, 19$, and 24, the model prediction is about to go to the wrong wing of the attractor, but EnKF updates the solution to the correct wing. The corresponding peaks in the estimated and actual errors indicate these unstable locations in state space. At $t = 5$ and 20, the estimated solution misses the reference solution's amplitudes, as reflected in the error estimates. Thus, the error-variance estimate is consistent, showing significant peaks at the locations where the ensemble has problems tracking the reference solution. Note also the similarity between the actual error and the estimated standard deviation. Thus, for all peaks in the residual, a corresponding one is present in the error variance estimate.

The error estimates show the same behavior as in Miller et al. (1994) with significant error growth when the model solution passes through the unstable regions of the state space, and otherwise weak error variance growth or even decay in the stable regions. For instance, observe the low error variance for $t \in [25, 28]$ corresponding to the solution's oscillation around one of the wings.

For this nonlinear problem, EnKF performs better than ES. *The reason is that the ensemble of realizations are recursively pulled toward the measured solution and are not allowed to diverge toward the wrong wing. In addition, the Gaussian update increments lead to an approximately Gaussian ensemble distributed around one of the wings. ES does not exploit this property of the sequential updating, and the realizations evolve freely and lead to non-Gaussian ensemble distributions.*

15.4 Ensemble Kalman Smoother Solution

The ensemble Kalman smoother (EnKS) is an extension of EnKF that allows for computing a smoother solution. In addition to updating the solution at the end of the time window, it uses time correlations from the ensemble to update the solution at all previously desired time instants. Thus, EnKS attempts to solve the recursive smoother problem from Sect. 2.4.3. It is a simple computation to obtain the EnKS solution for previous time-instants as soon as one has computed the EnKF solution. In the current example, the only code difference between the EnKF and EnKS is the number of time instants we include in the state vector. Also, when computing the EnKS solution, we define the number of previous time instants to update, and hence, we use a lagged EnKS implementation. The main additional computational cost of EnKS compared with EnKF is storing the ensemble of the variables we want to update at all time instants when we wish to compute the smoother solution. We refer to the extended discussion of EnKF and EnKS in Evensen (2009b, Chap. 9).

As for the EnKF solution we apply the Approxs. 1 (Markov model) and 2 (uncorrelated measurement errors in time). However, since the method allows for updating previous time windows, we do not apply the time window Approx. 3. The Approx. 4 (Gaussian priors) and 8 (the ensemble representation) apply as for EnKF. As when using ES and EnKF, the linearization, the sampling, and the linear regression approximations 5, 6, and 7 do not apply.

Figure 15.3 shows the solution obtained by EnKS. This solution is smoother than the EnKF solution and provides a better fit for the reference trajectory. In addition, EnKS recovers all of the problematic locations in the EnKF solution.

EnKS reduces the error estimates throughout the time interval, including the significant error peaks seen in the EnKF solution. As for the EnKF solution, there are corresponding peaks in the error estimates and the residuals, which suggests that the EnKS error estimate is consistent with the actual errors.

The code used in these examples is available from https://github.com/geirev/ EnKF_lorenz.git.

3Dvar and SC-4DVar for the Lorenz 63 Model

16

In this chapter, we study the workings of 3DVar and SC-4DVar on the same chaotic Lorenz 1963 system as used with ensemble methods in Chap. 15. We will apply both 3DVar and SC-4DVar sequentially over multiple data-assimilation windows, and we will demonstrate the difference between the filter solution obtained by 3DVar and the recursive SC-4DVar smoother solution. We will also dive deeper into the behavior of the SC-4DVar with highly nonlinear- and chaotic dynamics and try to understand more of the method's properties and possible limitations in these cases. After studying the 3DVar and 4DVar methods, we compare them with the ensemble methods used in Chap. 15.

16.1 Data Assimilation Set up

The governing equations of the Lorenz 1963 system are Eqs. (15.1)–(15.3) in Chap. 15. We use the standard parameter setting of $\sigma = 10$, $\rho = 28$, and $\beta = 3/8$, which leads to chaotic dynamics as depicted in Fig. 16.1. In all simulations the starting point for the true run is $(-10, -10, 20)^T$, and the time step $\Delta t = 0.01$.

We compute the background error covariance in all experiments by sampling a long run of the model every 16 time steps and calculating the sample covariance matrix. After that, we scale this matrix such that the maximum diagonal entry is 4, resulting in

$$\mathbf{C}_{xx} = \begin{pmatrix} 3.10839873 & 3.10666191 & -0.09539367 \\ 3.10666191 & 4.0 & -0.04713786 \\ -0.09539367 & -0.04713786 & 3.52161065 \end{pmatrix}.$$

Note the strong covariance between the x and the y components, related to the two wings in the x-y plane. The covariances with the z component are much smaller.

G. Evensen et al., *Data Assimilation Fundamentals*,
Springer Textbooks in Earth Sciences, Geography and Environment,
https://doi.org/10.1007/978-3-030-96709-3_16

Fig. 16.1 The plot illustrates the time evolution of the Lorenz 1963 system. Notice the two wings and the transition zone between these, which is mainly responsible for the chaotic dynamics, as seen by the spaghetti-like connections between the wings

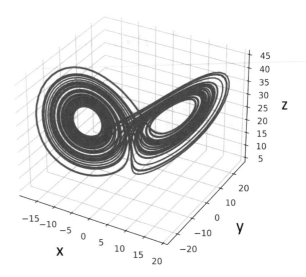

The z component has no knowledge of which of the wings or attractors the solution is on, see Fig. 16.1. From its construction, we can see that this covariance matrix contains the climatological correlations between the model variables and not the actual correlations at the start of a specific data assimilation experiment. It is a general weakness of a variational method that only tries to find the mode of the posterior pdf because one then ignores information on the uncertainty. The initial condition of the data-assimilation run is the true state at time zero perturbed by a random vector $\mathbf{C}_{xx}^{1/2}\boldsymbol{\xi}$ in which $\boldsymbol{\xi}$ is a random vector with elements drawn from a standard normal distribution.

We generate observations by sampling the true state at variable time intervals, with uncorrelated observation errors of standard deviation 1.0 and the identity measurement operator. We perform experiments where we observe either all variables or only the y component. When we only observe one variable, we illustrate how an SC-4DVar system (see Chap. 4) propagates or spreads the information from the measurement in time and among the variables.

In the following, we will run several experiments. We start by comparing SC-4DVar with 3DVar (see Chap. 6) to appreciate the strengths of a smoother, over a filter. After that, we will study the SC-4Dvar in more detail.

16.2 Comparing 3DVar and SC-4DVar

In this experiment, we run the system over 100 time steps with observations of all three variables only at the end of the assimilation window. We show the 3DVar solution in the upper half of Fig. 16.2. The black line is the true solution, and the red

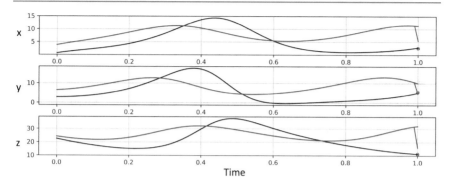

Fig. 16.2 The figure shows a typical 3Dvar solution x, y, and z. Red dots denote the observations for each variable at the end of the window, the black line is the true solution, the blue line the prior, and the purple line the analysis. Note that the 3DVar updates the model state only at the end of the window, and the blue and purple line are identical before the observation time

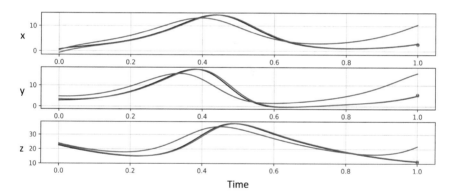

Fig. 16.3 A typical SC-4Dvar solution for x, y, and z. Red dots denote the observations for each variable at the end of the window. The black line is the true solution, the blue line the prior, and the purple line the analysis. In contrast to 3DVar, SC-4DVar updates the model trajectory over the whole assimilation window

dots at the end of the time window are the observations. The prior estimate is the blue line, while the purple line denotes the analysis trajectory. Note that the 3Dvar, only updates the solution at observation time, so the analysis only differs from the prior solution at the last time point.

In contrast, the SC-4DVar solution in Fig. 16.3 shows that the smoother solution updates the whole trajectory. The purple line is much closer to the truth at any time point. Note that the strong-constraint SC-4DVar used here only updates the initial condition at time zero and then uses the model to fill out the rest of the trajectory. Hence, the SC-4DVar scheme brings the information from the observation from the end of the assimilation window to the beginning. SC-4DVar computes this backward information propagation by solving the adjoint equations, as we have seen in Chap. 4.

16.3 Sensitivity to Observation Density in SC-4DVar

We will now examine the sensitivity of the SC-4DVar solution to the observation density in the assimilation window. We start by extending the assimilation window to 200 time steps, still only observing the state at the end of this window. As can be seen in Fig. 16.4 this is a challenging problem for SC-4DVar. In trying to fit the observations in the first two variables, the solution is worse for the third variable. The problem is complicated because the model trajectory passes through the unstable region where the two wings meet. In this region, the model evolution is very sensitive, and small perturbations will make the model solution go to one or the other wing. Since we assume no model errors, this strong sensitivity is carried over directly to the initial conditions, which the SC-4DVar is trying to estimate. This strong sensitivity manifests itself via multiple minima of the SC-4DVar cost function. We will elaborate on the appearance of local minima in the cost function, so multiple modes in the posterior pdf, in a later section. Finally, we should mention that if the truth stays in the stable regime in one of the winds of the attractor for a long time, the SC-4DVar can follow that solution over multiple oscillations.

The situation improves if we add more observations over the assimilation window. Figure 16.5 demonstrates that if we observe this system every 50 time steps, SC-4Dvar can find the initial condition that follows the truth quite well for 200 time steps. It means that the extra observations remove the multiple minima in the cost function, at least for the present prior initial conditions as we will elaborate on in Sect. 16.5.

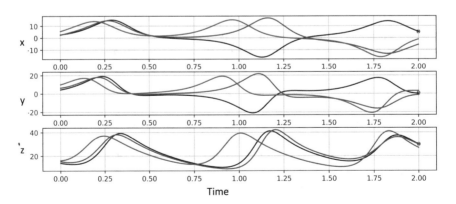

Fig. 16.4 Typical SC-4Dvar solution when the solution changes wings in the Lorenz 1963 system. Red dots denote the observations for each variable at the end of the window, the black line is the true solution, the blue line the prior, and the purple line the SC-4DVar analysis. Note that the SC-4DVar updates the model trajectory over the whole assimilation window but is unable to find the true trajectory

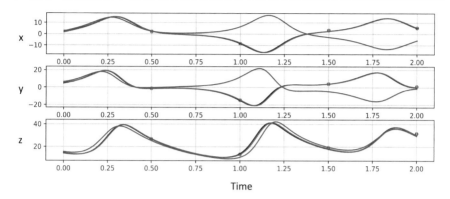

Fig. 16.5   Typical SC-4Dvar solution when the true trajectory changes wings in the Lorenz 1963 system, as in Fig. 16.4, but now with 4 times as many observations spread out over the assimilation window (red dots). The black line is the true solution, the blue line the prior, and the purple line the SC-4DVar analysis

16.4   3DVar and SC-4DVar with Partial Observations

We now run an experiment in which we only observe the y component and compare this to the case where we observe all three variables. We sample the observations every 100 time steps in a 200 timestep assimilation window to make this a challenge.

We show the results from only observing the y variable in Fig. 16.6. In contrast, the results from observing all three variables are indistinguishable from those displayed in Fig. 16.5, where we should remove the dots at 50 and 150 time steps. It is remarkable how well the SC-4DVar performs. To put this in perspective, we also compare these results with running 3DVar in this configuration in Fig. 16.7.

First, we notice that the 3DVar has to make a few strong adjustments to stay close to the true evolution of the system. The x variable is strongly updated in the right

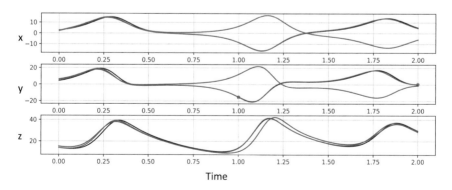

Fig. 16.6   4Dvar solution when the true trajectory changes wings in the Lorenz 1963 system with only the y variable observed at only 2 times in the assimilation window (red dots). The black line is the true solution, the blue line the prior, and the purple line the SC-4DVar analysis. Note the strong performance of the SC-4DVar

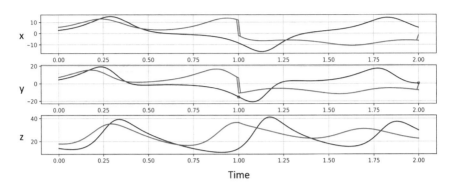

Fig. 16.7 3Dvar solution when the true trajectory changes wings in the Lorenz 1963 system with only the y variable observed at only 2 times in the assimilation window (red dots). The black line is the true solution, the blue line the prior, and the purple line the 3DVar analysis. Note the strong adjustments of the 3DVar at observation times

direction even though we only observe y, because of the 3DVar prior covariance matrix, which has a strong covariance between x and y, see Eq. (16.1). But the SC-4DVar is truly remarkable. It takes the influence from the observations of y at 100 and 200 time steps and brings those back to the initial condition at time zero via the adjoint equations. It updates all the variables and reruns the model over the window, providing perfect updates for x and z.

16.5 Sensitivity to the Length of Assimilation Window

A chaotic system such as Lorenz 1963 displays extreme sensitivity to small perturbations in initial conditions. (This extreme sensitivity is one of the definitions of chaos in the first place.) Many geophysical systems, such as the atmosphere, ocean, and climate systems, are chaotic, so the present experiments are so important. It will then come as no surprise that the SC-4DVar, which only updates the initial conditions, will be very sensitive to the nonlinearities in the likelihood, the actual realization of the measurement error, the data-assimilation window length, and the prior. We will discuss some of these sensitivities below.

It is well-known that the sensitivity grows with the length of the assimilation window, as we have seen in previous sections. Figure 16.8 shows the shape of the cost function plotted as a function of the x variable at the initial time for three different assimilation-window lengths. This cost function is the one that the SC-4DVar will minimize. The details depend on the prior initial condition and measurements and their error covariances, but the figure demonstrates the point. We see that, in this case, even for an assimilation window of length two non-dimensional time units, which corresponds to 200 time steps, multiple minima appear. And when the assimilation-window length increases to six, the cost function is very wild indeed, with hundreds of local minima.

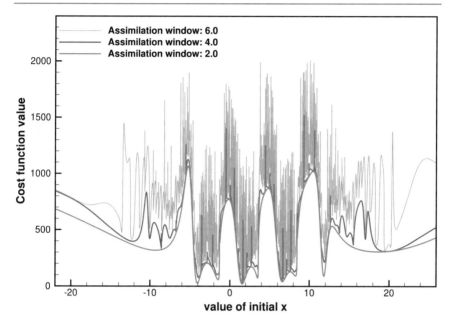

Fig. 16.8 Strong constraint penalty function for the Lorenz model as a function of the initial x-value, keeping y and z constant, when using data in the intervals $t \in [0, 2]$ (blue), $t \in [0, 4]$ (red), and $t \in [0, 6]$ (green), from Evensen (2009b)

The question then becomes how it is possible for SC-4DVar, which is a gradient method, to do a reasonable job on this system in the first place. The answer is threefold. First, the blue curve for the second assimilation window shows that the global minimum is at $x = 1.8$. If the prior mean would be close to that, say at $x = 2.8$, the SC-4DVar will find the global minimum. An initial error of the order of one, as in this case, or more minor, is not uncommon. However, if we would start with a similar initial error of one, but now at $x = 0.8$, the 4DVar solution would move off to the left, and it would not find the global minimum but the local minimum at $x = -1.3$. This discussion shows that the first guess, typically the prior mean, plays a significant role in chaotic systems.

Another reason for the excellent performance of SC-4DVar is that Fig. 16.8 does not show full cost function in the 3-dimensional solution space. The local minima shown in this cross-section may be connected by "valleys" in 3-dimensional space. In that case, the local minima displayed here might not be actual local minima. Of course, with longer assimilation windows such as four and six, it is implausible that there will not be many local minima.

The final reason why 4DVar often gives reasonable answers is the width of the prior pdf. In Fig. 16.8 the prior was relatively wide. One can imagine that a much narrower prior will smooth out the cost function because we can not reach many of the local minima in the set assimilation-window length. A clear example comes from numerical weather prediction. ECMWF is the only operational center that performs an SC-4DVar over a 12-hour window, while the other centers use 6-hour windows.

The reason is twofold, the highly accurate prior in the 4DVar system and the superior treatment of the complex satellite observations.

One could now pose the question why we want to run long-window SC-4DVar. The main reasons are the accurate covariance between variables and the number of observations. A longer assimilation window means that we use more future measurements to find the best estimate. Remember that the SC-4DVar uses the adjoint technique, which tells us how small perturbations around a fully nonlinear model trajectory move through the assimilation window. Hence, we can interpret this as having a space-time covariance matrix around a fully nonlinear model trajectory. This space-time covariance matrix adapts to the system's local space-time dynamics through its connection with the nonlinear model trajectory. As a result, as long as the perturbations remain small, this *implicit* prior covariance matrix over the whole assimilation window is superior to anything we can generate otherwise. For instance, in any ensemble smoother, we would typically need the localization of the ensemble space-time covariance matrix, which we avoid here.

However, it is good to remember that the above statement is only valid when the linearization is accurate, which means when the prior remains close enough to the true system, both in terms of mean and a small uncertainty. In a chaotic system, the time window that remains valid is always finite, and the only fundamental way to avoid this issue is to include model errors. Model errors allow for adaptation of the model trajectory within the assimilation window, removing part of the sensitivity to the initial conditions.

A practical way to partly remove the sensitivity of the initial conditions on the assimilation window is to divide the assimilation window up into smaller pieces. We first run an SC-4DVar over the first piece, likely resulting in a reasonable initial condition estimate because of the shorter window length. Then this more accurate estimate is used as the first guess in a window that is twice longer. Since the first guess is better, we again can assume a more precise 4DVar solution over this longer window. We can repeat this procedure to cover the whole original assimilation window finally. This idea by Pires et al. (1996) significantly improves the results of SC-4DVar in small-dimensional and highly nonlinear systems.

Another practical solution employed by many operational weather prediction centers is to run the initial outer loop iterations of the SC-4DVar with a reduced model resolution and a reduced observation set. The resolution and observation density increase in later outer loop iterations, slowly bringing in more nonlinearity while starting each outer-loop iteration from a more accurate first guess. This approach leads to a much more linear data-assimilation problem that is suitable for SC-4DVar.

16.6   SC-4DVar with Multiple Assimilation Windows

We will now show results from running SC-4DVar over multiple data assimilation windows. Figure 16.9 shows results from 10 assimilation windows observing all three variables every 50 time steps in an assimilation window of 200 time steps.

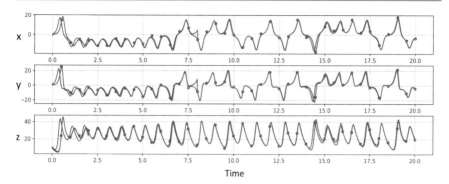

Fig. 16.9 SC-4DVar results over 10 assimilation windows of 200 time steps each. Truth (black), prior (blue) and analysis (purple). The solution is very accurate, and the black line is almost completely covered by the purple line

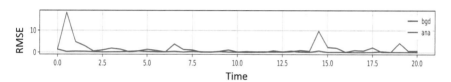

Fig. 16.10 RMSE versus time for SC-4DVar run over 10 assimilation windows. The blue line is the background error and the purple line is the analysis error

With this observation frequency observing only the y variable did not work well after two assimilation windows, showing that one has to be careful with results in short data-assimilation experiments.

Figure 16.10 demonstrates that the analysis's root-mean-square error (RMSE) is tiny, of the order of the observational error of 1, as expected. The background error fluctuates dramatically around 100 and 1400 time steps, typically related to the end of a forecast window.

Finally, we produce a zoom in on the solution when we observe only variable y at a transition between 2 assimilation windows in Fig. 16.11. We see that even the analysis is not smooth at this transition. This result should not come as a surprise because the minimizations at each side of the transition see a different set of observations. Such jumps are standard in strong-constraint SC-4Dvar solutions and are present in atmospheric reanalyses. They arise because one cannot run a SC-4DVar over a too-long window because of the sensitivity to initial conditions in a chaotic system such as the atmosphere or the ocean. To avoid these jumps at the end of assimilation windows, one would have to run a weak-constraint SC-4DVar, in which the influence of the initial conditions becomes negligible after some time. Indeed, the oceanographic ECCO system does provide smoother solutions with assimilation windows of 50 years or more, albeit at relatively low spatial resolution.

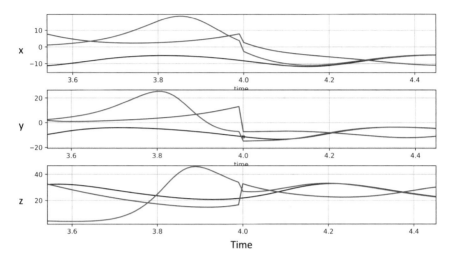

Fig. 16.11 Zoom in of the analysis solution at the boundary of two assimilation windows. Note that the total solution is not smooth in time because the two solutions on either side of time $t = 4$, corresponding to 400 time steps, are from two independent minimizations

16.7   A Comparison with Ensemble Methods

Finally, we make a comparison with ensemble methods. We can directly compare the 3DVar to an EnKF. The main difference is that the 3DVar uses a climatological prior covariance matrix, and the EnKF an ensemble-based dynamical prior covariance matrix at each observation time. If the ensemble is large enough, the latter will be more accurate as it contains flow information, while the former does not. As an example, just after the solution passes the transition region where the decision on which wing the system will be in, the uncertainty is high, and the EnKF ensemble spread is large, see the red line in the lower figure in Fig. 15.2. The enhanced uncertainty allows for the following observation to firmly pull the ensemble to the correct wing, as seen in the upper panel of that figure.

In contrast, when the solution circles around in one of the wings, the error growth is small, the ensemble spread remains small, demonstrating that the uncertainty in the solution is small. In contrast, the 3DVar prior covariance matrix is some average of these situations. The climatological variance of the x variable is 3.1, while, in the EnKF, it is of order 1.5. We see that the 3DVar covariance values are quite large and hence conservative. This situation is similar for a 4DVar, and neither of these methods will perform well without these conservative prior covariances.

We should compare 4DVar with the ES and the EnKS or their iterative variants. One apparent issue is that the SC-4DVar struggles with window lengths larger than two, so 200 time steps, while in the previous chapter, we saw in Fig 15.1 that the ES manages to follow the truth quite well, placing the solution in the right wing all the time, for 4000 timesteps! The ES solution was imperfect, and analysis errors were significant but still consistent with the ensemble spread. The main difference is that the SC-4DVar only adjusts the initial conditions, while the ES updates the whole model trajectory at every time step. Hence, the ES can follow the observations quite well.

The EnKS and the SC-4DVar perform quite similarly as long as the assimilation windows for the latter are not too long, typically two non-dimensional time units or shorter for the Lorenz 1963 system. Since SC-4DVar decouples the analysis windows, jumps will occur between assimilation windows, as shown in Fig. 16.11. Inside a window, the solution is perfectly smooth, following the model equations exactly. In contrast, the EnKS update also affects the previous assimilation windows, and we obtain a smooth solution from one assimilation window to the next, but the model equations are only followed exactly between updates.

Representer Method with an Ekman-Flow Model

17

Eknes and Evensen (1997) solved the weak-constraint variational problem for a linear Ekman-flow model using the representer method. They computed the weak constraint solution for a long time series of velocity measurements. Additionally, they considered a parameter-estimation problem which rendered the problem nonlinear. Here we will focus on the representer method's properties for a linear problem. The model is simple and allows for a straightforward interpretation and demonstration of the method. For more details, we refer to Eknes and Evensen (1997) and Evensen (2009b).

17.1 Ekman-Flow Model

The Ekman-flow model describes the horizontal velocity field as a function of depth in the ocean's upper surface layer when subject to a wind forcing. The model provides the so-called Ekman spiral due to the Coriolis force and we can write its equations in a nondimensional form as

$$\frac{\partial \mathbf{u}}{\partial t} + \mathbf{k} \times \mathbf{u} = \frac{\partial}{\partial z}\left(A \frac{\partial \mathbf{u}}{\partial z}\right), \qquad (17.1)$$

where $\mathbf{u}(z, t)$ is the horizontal velocity vector, $\mathbf{k}$ is a vertically pointing unit vector, and $A = A(z)$ is the diffusion coefficient. The initial conditions are

$$\mathbf{u}(z, 0) = \mathbf{u}_0, \qquad (17.2)$$

and we specify boundary conditions as a wind-drag at the surface $z = 0$ and zero drag or no friction at the lower boundary $z = -H$ as

© The Author(s) 2022

G. Evensen et al., *Data Assimilation Fundamentals*,
Springer Textbooks in Earth Sciences, Geography and Environment,
https://doi.org/10.1007/978-3-030-96709-3_17

$$A \frac{\partial \mathbf{u}}{\partial z}\bigg|_{z=0} = \left(c_d \sqrt{u_a^2 + v_a^2} \right) \mathbf{u}_a, \tag{17.3}$$

$$A \frac{\partial \mathbf{u}}{\partial z}\bigg|_{z=-H} = \mathbf{0}, \tag{17.4}$$

where c_d is the wind drag coefficient, and $\mathbf{u}_a$ is the atmospheric wind speed. Following the procedure outlined in Sect. 5.5, we can derive the Euler–Lagrange equations and the representer solution. We refer to the original work by Eknes and Evensen (1997) for the detailed derivation.

17.2 Example Experiment

We now discuss a simple example to illustrate the representer method. We use a constant wind with $\mathbf{u}_a = (10\,\mathrm{m\,s}^{-1}, 10\,\mathrm{m\,s}^{-1})$ to spin up the velocity structure in the first-guess solution, which we initialize with $\mathbf{u}(z, 0) = \mathbf{0}$ and then perform 50 h of integration. We construct the reference case and extract velocity data by continuing the integration for another 50 h.

By measuring the reference case and adding Gaussian noise, we generate nine simulated measurements of $\mathbf{u}$; i.e., we have a total of 18 measurements of the u and v components of the velocity at three different depths. Figure 17.1 illustrates the measurement locations, together with the first-guess, the reference solution, and the posterior mode from the representer method. The reference solution is regenerated quite well, even though the first-guess solution is out of phase with the reference case, and the measurements do not resolve the period of the oscillation. A single measurement may suffice for reconstructing the correct phase since the corresponding representer will carry the information both forward and backward in time. However, the errors will be more significant with fewer measurements.

To illustrate the solution procedure using the representer method in more detail, Fig. 17.2 presents the u-components of the variables $\mathbf{s}_5$, $\mathbf{r}_5$, λ, and the convolutions $\mathbf{C}_{qq} \bullet \mathbf{s}_5$ and $\mathbf{C}_{qq} \bullet \lambda$. The $\bullet$ denote convolution in space and time, but we set the model errors' time correlation to zero in this example. The symbols follow the notation of Sect. 5.5. The subscript five denotes the measurement number. Measurement number five corresponds to the u component at the location $(z, t) = (-20.0, 25.0)$. These plots demonstrate how the information from the measurements influences the solution.

The upper plot shows the u-component of adjoint representer $\mathbf{s}_5$ forced by the impulse function at the measurement location, see Eq. (5.41). This information is then propagated backward in time while the u and v adjoint representer velocity components interact during the integration.

After that, we use $\mathbf{s}_5$ on the right-hand side of the forward Eqs. (5.38) and (5.39) to evaluate the representer's initial and boundary conditions and forcing fields. The convolution $\mathbf{C}_{qq} \bullet \mathbf{s}_5$, is a smoothing of $\mathbf{s}_5$ according to the covariance functions contained in $\mathbf{C}_{qq}$, as seen from the second plot in Fig. 17.2.

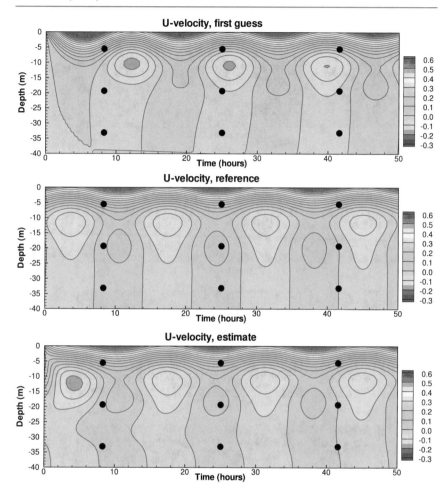

Fig. 17.1 The u components of *(from top to bottom)* the first-guess estimate $\mathbf{u}_F$, the reference case $\mathbf{u}$ and the posterior mode solution for $\mathbf{u}$. The contour intervals are $0.05\,\mathrm{m\,s^{-1}}$ for all the velocity plots. The measurement locations are marked with a bullet. The v components are similar in structure and not shown. Reproduced from Eknes and Evensen (1997)

The representer $\mathbf{r}_5$ is smooth and is oscillating in time with a period reflecting the inertial oscillations described by the dynamical model. Note that the representers will have a discontinuous time derivative at the measurement location since the right-hand side $\mathbf{C}_{qq} \bullet \mathbf{s}_5$ is discontinuous there. However, if we had included a time-correlation in $\mathbf{C}_{qq}$, then $\mathbf{C}_{qq} \bullet \mathbf{s}_5$ would be continuous, and the representer $\mathbf{r}_5$ would be smooth.

After computing and measuring the representers of all observations to generate the representer matrix $\mathcal{R}$, we solve for the vector $\mathbf{b}$ from Eq. (5.42). We then use $\mathbf{b}$ in (5.43) to decouple the Euler–Lagrange equations. The u-component of $\boldsymbol{\lambda}$ (Fig. 17.2) illustrates how the various measurements have a different impact determined by

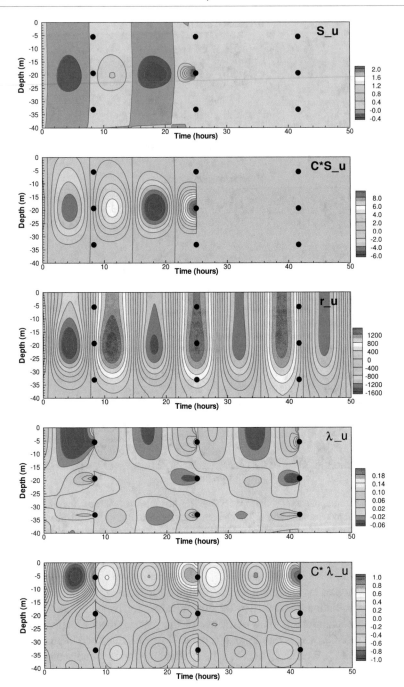

Fig. 17.2 The u component of *(top to bottom)* $\mathbf{s}_5$, $\mathbf{C}_{qq} \bullet \mathbf{s}_5$, $\mathbf{r}_5$, the adjoint λ, and $\mathbf{C}_{qq} \bullet \lambda$. The measurement locations are marked with a bullet. The v components are similar in structure and not shown. Reproduced from Eknes and Evensen (1997)

values of the coefficients in **b**. After solving for λ, we construct the right-hand side in the forward model equation through the convolution $\mathbf{C}_{qq} \bullet \lambda$, given at the bottom of Fig. 17.2. The role of this term is to force the solution to smooth the measurements.

17.3   Assimilation of Real Measurements

We will now apply the representer method with the LOTUS–3 data set (Bowers et al., 1986) in a similar setup to the one used by Yu and O'Brien (1991, 1992). The LOTUS–3 experiment sampled measurements in the northwestern Sargasso Sea (34° N, 70° W) during summer 1982. Current meters at depths of 5, 10, 15, 20, 25, 35, 50, 65, 75, and 100 m measured the currents. A wind recorder mounted on top of the LOTUS–3 tower measured the wind speed. The sampling interval was 15 min, and we are using data from June 30 to July 9, 1982. We have further subsampled the data every five hours at the depths 5, 25, 35, 50, and 75 m. The reason for not using all the measurements is to reduce the size of the representer matrix. The data still resolve the inertial period and the vertical length, and we expect only to ignore small-scale noise by the subsampling.

We initialized the model from the first measurements collected on June 30, 1982. The standard deviation of the small-scale variability of the velocity observations was close to $0.025 \, \mathrm{m \, s}^{-1}$, and we used this value to determine the error variances for the observations and the initial conditions. We specified the model error variance after a few runs to give a relatively smooth posterior mode estimate. We want a solution that nearly satisfies the model equations and is close to the observations without over-fitting them.

Figure 17.3 shows the results from the estimation as time series of the u component of the velocity at various depths. The figure plots the posterior-mode estimate together with the complete time series of measurements. We denote the measurements used in the estimation as bullets. The estimate's amplitude and phase agree well with the measurements at all times and depths. Note also that the estimate is smooth and does not precisely interpolate the measurements. By a closer examination of the solution, it is possible to see that the time derivative is discontinuous at measurement locations due to neglecting the time correlation in the model error covariances.

For comparison, we present a strong-constraint solution in Fig. 17.4. It is clear from comparisons that the strong-constraint solution, as determined by the initial conditions, in the upper part of the ocean is reasonably in phase with the measurements. At the same time, the amplitudes are not as good as in the weak-constraint solution. The only way the amplitudes can change when the model is assumed to be perfect is by vertical transfer of momentum from the surface. Thus, we obtain a good fit near the surface, while there is hardly any effect from the wind stress at depth, and the strong-constraint solution is also far from the measurements. The strong-constraint estimate is close to a sine curve representing the model's inertial oscillations. These results indicate that model deficiencies, such as neglected physics,

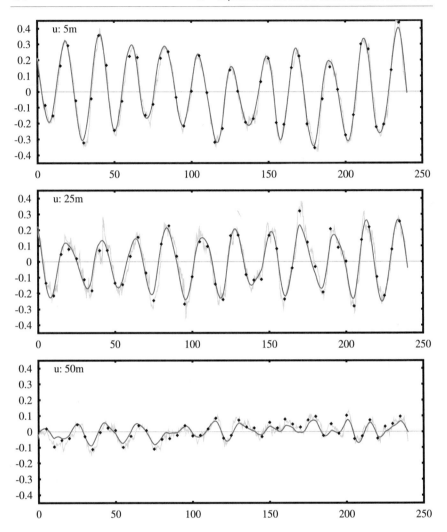

Fig. 17.3 Weak-constraint results from the LOTUS–3 assimilation experiment from Eknes and Evensen (1997). Inverse estimate for the u component of velocity *(red lines)*, the time series of measurements *(blue lines)*, and the subsampled measurements *(bullets)*, at 5, 25 and 50 m

should be accounted for through a weak-constraint variational formulation to ensure the solution agrees with the measurements.

The representer solution provides the optimal minimizing solution of the linear inverse problem. Furthermore, the M-dimensional space spanned by the representers contains the optimal solution update. Note that the equation for **b**, (5.42) is similar to the one solved in the analysis scheme in the standard Kalman filter. The representers correspond to the measurements of the space-time error covariance of the first-guess

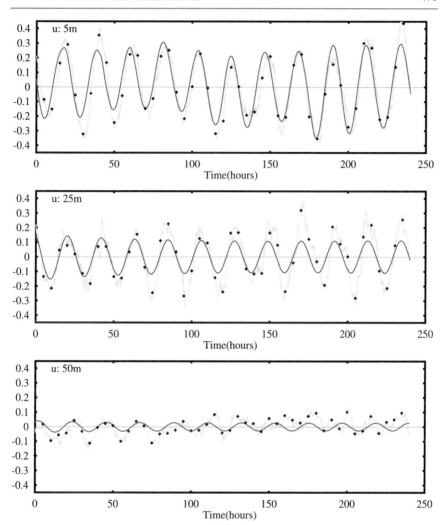

Fig. 17.4 Strong-constraint results from the LOTUS–3 assimilation experiment from Eknes and Evensen (1997). Inverse estimate for the u component of velocity *(red lines)*, the time series of measurements *(blue lines)*, and the subsampled measurements *(bullets)*, at 5, 25 and 50 m

solution. Thus, there are strong similarities between the analysis step in the ensemble Kalman smoother and in the representer method. To summarize, the representer method is a highly efficient approach for solving linear inverse problems, and it is also applicable to many nonlinear dynamical models.

Comparison of Methods on a Scalar Model

18

This chapter aims to demonstrate the impact of some of the critical approximations we apply with different assimilation methods. Using a simple scalar model, we can simulate many realizations and eliminate sampling errors. We will examine the advanced particle flow and compare its performance to iterative smoothers and the linear updates from the ensemble Kalman filter (EnKF) or ensemble smoother (ES). By testing the methods on models of varying degrees of nonlinearity, we develop an overall understanding of how different data-assimilation methods perform in different situations.

18.1 Scalar Model and Inverse Problem

Let's define two nonlinear scalar models

$$y = g(x, q) = x + \beta x^3 + q, \tag{18.1}$$

and

$$y = g(x, q) = 1 + \sin(x) + q, \tag{18.2}$$

which, given an initial state x and a model error q, define a prediction y.

In Eq. (18.1) β is a parameter that determines the non-linearity of the model. In the current example, we have used $\beta = 0.3$. Evensen (2018) used the same model but without the model error, while Evensen (2019) included model errors but used $\beta = 0.2$. This model introduces a nonlinearity while retaining a monotonic model response given the inputs x and q. In Eq. (18.2), we introduce a model with stronger nonlinearity, resulting in multimodal posteriors, since one model output can result from different model inputs.

The goal is to demonstrate the impact of some of the approximations we introduced in Chap. 2 when we try to sample the Bayes' posterior for x and q given a

© The Author(s) 2022
G. Evensen et al., *Data Assimilation Fundamentals*,
Springer Textbooks in Earth Sciences, Geography and Environment,
https://doi.org/10.1007/978-3-030-96709-3_18

measurement of y. We will use techniques that solve the data-assimilation problem with different approximations. We start with a particle-flow approach that samples the Bayes' posterior exactly when we use an adjoint-based model sensitivity. Then, we introduce the approximate ensemble-based model sensitivity in Approx. 7 to examine its impact. The EnRML approach with adjoint model sensitivity minimizes the cost functions Eq. (7.1) exactly. Still, this approach only approximately samples the Bayes' posterior due to Approx. 6. We further examine the impact of the linearization in Approx. 5 introduced by the ES scheme. Finally, we examine the convergence of the ESMDA method.

18.2 Discussion of Data-Assimilation Examples

We run three cases of different levels of nonlinearity. In the first case, we use the model in Eq. (18.1). We sample the prior ensemble x_j^{f} from a normal distribution $\mathcal{N}(x^{\mathrm{f}} = 0.0, C_{xx} = 1.0)$. and we sample the perturbed observations, d_j of y, from $\mathcal{N}(0.0, 1.0)$. Thus, except for the model nonlinearity, this is a rather trivial case.

In the second case, we also use the model in Eq. (18.1), but now we sample the prior from to $\mathcal{N}(1.0, 1.0)$ the perturbed measurements from $\mathcal{N}(-1.0, 1.0)$. Thus, we introduce stronger nonlinearity and non-symmetry to the problem.

In the third case, we use the multimodal model in Eq. (18.2) with the prior ensemble sampled from $\mathcal{N}(1.0, 1.0)$ and the measurement ensemble from $\mathcal{N}(0.0, 1.0)$. The model error q is a random variable sampled from $\mathcal{N}(0, C_{qq} = 0.25)$ for all the three cases.

In these examples, we use a sufficiently large number of samples, i.e., 10^7, to generate accurate estimates of the probability density functions. Furthermore, the large ensemble allows us to work directly with the pdfs and examine the converged solutions of the methods and the impact of the various approximations. We use a smaller ensemble of 10^5 realizations for the particle flow computations due to the increased computational cost of the kernel matrix multiplications.

Stordal et al. (2021) discussed the particle flow for this type of data-assimilation problem. Given a cost function as in Eq. (3.9) and its gradient Eq. (3.10) we can write an iteration using the gradient from Eq. (9.49) as

$$\mathbf{z}_j^{i+1} = \mathbf{z}_j^i - \gamma \frac{1}{N} \sum_{l=1}^{N} \mathbf{K}(\mathbf{z}_l^i, \mathbf{z}_j^i) \nabla_{\mathbf{z}_l^i} \mathcal{J}(\mathbf{z}_l^i) - \nabla_{\mathbf{z}_l^i} \mathbf{K}(\mathbf{z}_l^i, \mathbf{z}_j^i). \tag{18.3}$$

Here we use the particle-flow filter from Eq. (9.49) and that we can write any posterior pdf as $f(\mathbf{z}|\mathbf{d}) \propto \exp\left(-\mathcal{J}(\mathbf{z})\right)$, such that the gradient of the log-posterior is identical to the gradient of the cost function $\mathcal{J}(\mathbf{z})$. As explained in Sect. 9.3.2, we can choose any symmetric smooth kernel when the number of particles is large, and in this example, we use a Gaussian kernel

$$\mathbf{K}(\mathbf{z}_l^i, \mathbf{z}_j^i) = \exp\left(-\left(\mathbf{z}_l^i - \mathbf{z}_j^i\right)^{\mathsf{T}} \mathbf{C}^{-1}\left(\mathbf{z}_l^i - \mathbf{z}_j^i\right)\right). \tag{18.4}$$

Using a "narrow" kernel will lead to an ineffective repulsion term and a gradient term where only realization j impacts the gradient for itself. In the examples below, we used a diagonal C with the number $2/\pi$ on the diagonal.

In Figs. 18.1, 18.2 and 18.3 we present the results from the three cases with increasing degrees of nonlinearity. From top to bottom, we show the results using particle flow, an iterative ensemble smoother (IES) based on EnRML, ES, and finally, ESMDA.

In the upper plots we see that the particle-flow method with adjoint model sensitivity (full line) converges to the correct posterior independent of model nonlinearity. Moreover, the technique even recovers the bimodality in the rightmost plot. This result is in agreement with the theory. When introducing the ensemble-averaged model sensitivity from Approx. 7 in the particle flow methods, we still obtain good results in the weakly nonlinear cases with only a slight distortion of the sampled distribution compared to the true posterior. However, in the bimodal case, none of the methods show any skill in recovering the correct distribution. Thus, the linear regression Approx. 7 requires the model to be weakly nonlinear in the sense that the model is a monotonic function of the estimated inputs.

Although the particle-flow algorithm shows excellent potential for sampling the posterior pdf, extending it to higher-dimensional problems and practical applications poses certain challenges. For example, in a high-dimensional model, we must ensure that the repulsion term with a given kernel is "active," using an appropriate kernel and a sufficient number of samples.

In the second row of plots in Figs. 18.1, 18.2 and 18.3 we computed the results from the EnRML sampling where we apply the Approx. 6 and minimize an ensemble of cost functions Eq. (7.1) using the IES. The solid line shows the sampled pdf when using the adjoint model sensitivity, while the dashed lines illustrate the results when we introduce the ensemble-averaged model sensitivity from Approx. 7. In the nearly linear case in Fig. 18.1, the EnRML sampling using both the adjoint and the linear regression representation for model sensitivity gives nearly the same answer. However, when the degree of nonlinearity increases, the two estimates diverge from each other and the true posterior. In the bimodal case, the EnRML sampling with adjoint model sensitivity captures both the modes but only approximately. This example represents all methods that exactly minimize the cost functions in Eq. (7.1), including the En4DVar.

The third row of plots in Figs. 18.1, 18.2 and 18.3 shows the results when we minimize the cost functions in Eq. (7.1) using the ES method that includes the additional linearization of Approx. 5. Clearly, in both cases using an ensemble-averaged linear regression and an adjoint model sensitivity, the linearization causes the resulting pdf to deviate even further from the true posterior pdf. Surprisingly, we obtain a nearly perfect fit using ES in the case with the multimodal model when using the adjoint model sensitivity. In these plots, we also show the result with an ensemble-averaged sensitivity and where we computed $\mathbf{C}_{yy}$ directly without using the correct regression in Eq. (7.11). This error causes a significant shift in the estimated pdf and a worsening of the result.

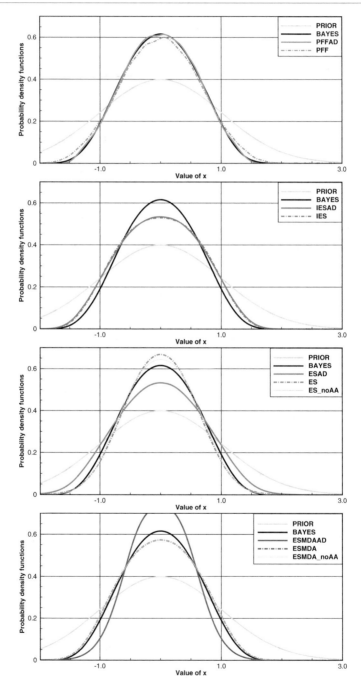

Fig. 18.1 Sampling Bayes' for the scalar inverse problem in Eq. (18.1) with the prior and observation centered at zero. From top to bottom we show results using particle flow (PFF), an iterative ensemble smoother (IES), ES, and ESMDA. Each plot shows the results using an ensemble based model sensitivity (full lines) and an adjoint model sensitivity (dashed lines). Additionally, for ES and ESMDA, we plot the results when omitting the correction when computing $\mathbf{C}_{yy}$

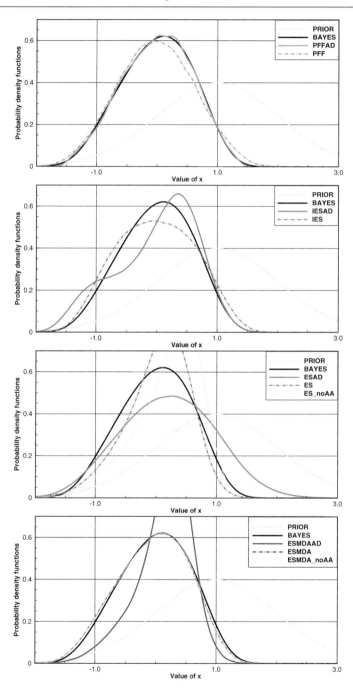

Fig. 18.2 As in Fig. 18.1 but with the prior centered at $x = 1$ and the measurement at $x = -1$

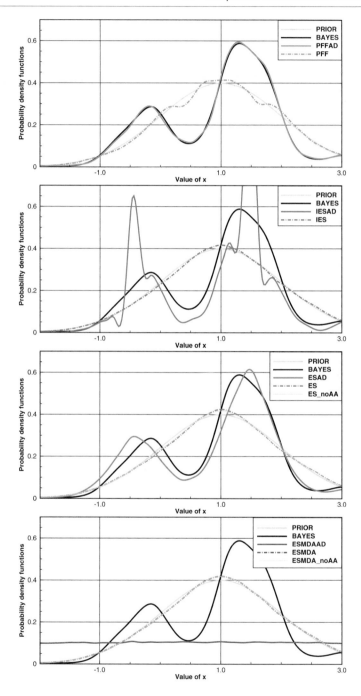

Fig. 18.3 As in Fig. 18.1 but for the highly nonlinear model defined by Eq. (18.2)

Finally, the bottom plots of Figs. 18.1, 18.2 and 18.3 show the results when sampling the posterior distribution using ESMDA. Recall that ESMDA is just a sequence of ES updates using measurements with inflated errors. In the linear case, ESMDA samples exactly the Bayes' posterior pdf. In the nonlinear case, ESMDA is subject to the linearity Approx. 5, but the impact of this approximation can be made negligible by using many small update steps. The many minor updates are nearly linear, and we have an almost insignificant effect of using the regression formula in Eq. (7.11). ESMDA with an ensemble gradient does amazingly well for this particular problem, while when using the adjoint model sensitivity, we obtain a much worse result. In the highly nonlinear case in Fig. 18.3, we observe that ESMDA with 32 steps and an adjoint model sensitivity does not work. This problem is a result of the vastly inflated measurements used in each update step. The large perturbations create some perturbed measurements with non-physical values that exceed the possible function outputs. For the model in Eq. (18.2), the outputs are restricted to the interval $y \in [0 : 2]$, neglecting the stochastic model errors q. Evensen (2018) pointed out this issue with ESMDA. Using fewer update steps or even a square root formulation for the ESMDA (Emerick, 2018) might resolve this problem.

In summary, the particle-flow algorithm converges to the Bayesian posterior when using the adjoint model sensitivity. However, the computational cost is significantly larger than for the other methods. Moreover, we must choose the kernel wisely and care about the required number of particles for high-dimensional systems. Furthermore, introducing an ensemble gradient results in a solution that diverts from the correct Bayesian solution. Thus, when we don't have an adjoint, we may still obtain an improved estimate using particle flow. Finally, the EnRML solution introduces another approximation, and the result is less good than the particle-flow solution. The code used for these examples is available from https://github.com/geirev/EnKF_scalar.git.

18.3 Summary

As we have seen from the examples discussed in this chapter, the approximate sampling of the Bayes' posterior worsen with increasing nonlinearity. Particularly for the multimodal problem, the results from using ensemble smoothers with an ensemble model sensitivity completely fail. In Fig. 18.4 we illustrate the basis for these methods. In this example, we have a monotonically increasing prediction with an increasing value for the input parameter. This situation corresponds to the two examples in Figs. 18.1 and 18.2. Thus, we have a positive correlation between the input parameter and the predicted measurement. When we are underpredicting the observation, the positive correlation indicates that we can increase the input-parameter value to obtain a better fit. The ES is using precisely this approach. The iterative EnRML

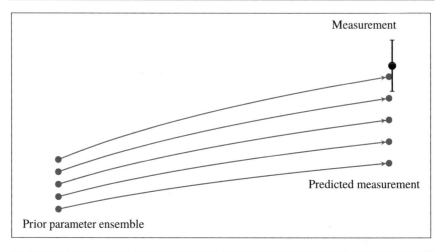

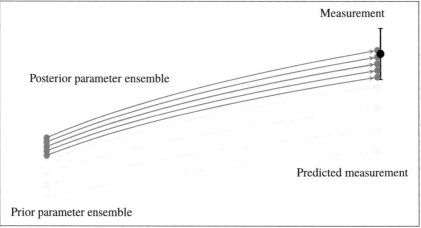

Fig. 18.4 Illustration of the linear regression update used to solve the parameter estimation problem. The upper frame illustrates the ensemble prediction of five realizations of a scalar parameter, which misses the measurement in blue. The lower frame shows how the updated parameter realizations lead to predictions in better agreement with the measurement

smoothers take this one step further to handle better a certain degree of nonlinearity in the dynamics. However, the stronger the nonlinearity, the poorer is the regression representation of the model sensitivity. And, for non-monotonic functions, the correlation between input parameters and predicted measurements approach zero, explaining the poor performance of the regression updates in Fig. 18.3.

Particle Filter for Seismic-Cycle Estimation

19

The particle filter is an effective data-assimilation method for low-dimensional, non-linear systems. It is easy to implement, and it is straightforward to include model error, parameters, and controls in the state vector. This chapter demonstrates the use of a particle filter in the case of a parameter bias. We apply the standard particle filter with importance-resampling to estimate seismic cycles. Seismic-cycle estimation helps us assess and forecast fault slips in slow-slip events and earthquakes. In the simulation of seismic cycles, the parameter choice determines the periodicity of the seismic events. Consequently, having a parameter bias will result in errors in the simulated state trajectory. This chapter illustrates how we can compensate for errors in the state trajectory caused by a parameter bias using the particle filter. We include the model errors, and the joint estimation of state and parameters leads to more satisfactory results than we can obtain by estimating the state alone. We compare state-estimation results between the standard particle filter and EnKF. Furthermore, we illustrate the advantage of using a fully nonlinear data-assimilation method for systems with sudden transitions in the model trajectory.

19.1 Particle Filter for State and Parameter Estimation

We apply the particle filter with sequential importance resampling as described in Sect. 9.2.1 on an application for seismic-cycle estimation. This application provides an interesting data assimilation case as the state trajectory is sensitive to both parameters and state, which are generally uncertain.

The highly nonlinear behavior of seismic cycle models may limit the effectiveness of the EnKF or related methods. Within a set of realizations for seismic cycles, seismic events can occur at different times for each of the realizations. In this case,

© The Author(s) 2022
G. Evensen et al., *Data Assimilation Fundamentals*,
Springer Textbooks in Earth Sciences, Geography and Environment,
https://doi.org/10.1007/978-3-030-96709-3_19

the particle filter has the advantage of propagating the full pdf of the state. The standard particle filter applied in this chapter only affects the weight of the particles. As we will illustrate for a quasi-geostropic ocean model in Chap. 20, it will not move the particles as the particle flow filter does.

19.2 Seismic Cycle Model

To simulate the stick-slip motion of a fault related to the movement of a tectonic plate, Burridge and Knopoff (1967) introduced a simplified representation of an assembly of springs and blocks. This Burridge–Knopoff (BK) model connects a set of blocks through springs. It also uses springs to connect the blocks to a loader plate. The loader plate moves with a uniform velocity and pulls the blocks. Here, we consider a zero-dimensional version of the BK model that represents a single spring-block system, as shown in Fig. 19.1. The contact surface, represented by the block and the rough surface over which the block can slide, simulates the fault.

The equations for this rate- and state-dependent friction law can be found in Ruina (1983). Erickson et al. (2011) shows how to solve these equations by rewriting them as three partial differential equations

$$\frac{\partial \theta}{\partial t} = -v(\theta + (1 + \zeta) \ln v), \tag{19.1}$$

$$\frac{\partial u}{\partial t} = v - 1, \tag{19.2}$$

$$\frac{\partial v}{\partial t} = -F^2 \left(u + \frac{1}{\xi}(\theta + \ln v) \right), \tag{19.3}$$

where F is the non-dimensional frequency of the resulting cyclic motion, and ξ is a non-dimensional spring constant. Furthermore, u is the block's non-dimensional

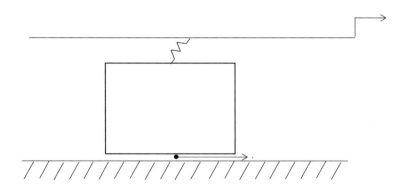

Fig. 19.1 The figure illustrates a zero-dimensional version of the spring-block system by Burridge and Knopoff (1967), consisting of a block connected with a spring to a driving plate moving with a constant velocity

slip, i.e., displacement, relative to the driver plate, and v is the corresponding slip rate. Finally, θ characterizes the state of the frictional surface. We interpret it as a measure of the average lifetime of the asperity contact population in a fault, a variable that influences the strength of the contacts. We obtain the fault shear stress τ can by multiplying the slip values with $-\xi$. Banerjee et al. (2022) give further details on the parameters used in the simulations.

The parameter ζ is the parameter of interest here. It depends on the friction parameters in the rate- and state-dependent friction law and determines the stick-slip behavior of the slip events. In our zero-dimensional model, we choose the parameter ζ such, that the system has a stable limit cycle solution with periodic oscillations. We investigate the effectiveness of the particle filter in a case of a biased ζ. A bias in ζ affects the simulated amplitude and frequency of fault stress, slip and the state variable θ in the earthquake cycle.

19.3   Data-Assimilation Experiments

We apply a standard particle filter with sequential importance resampling in three cases:

Case A: state estimation,
Case B: state estimation with increased model error,
Case C: state- and parameter estimation.

In all three cases, we represent the model noise by an additive model error dq in Eq. 2.24). We add this model error to the Eq. (19.2) for $\frac{\partial u}{\partial t}$. For Case A and Case C, we draw the model error from a Gaussian distribution with zero mean and a standard deviation of 0.01. For Case B, the standard deviation of the model error is 0.1.

We run the forward model for a period of 500 time units. We define the system's state by the fault shear stress τ, the slip rate v, and the state variable θ at the fault interface. In Case A and Case B, the state vector contains the model state plus the model error $\mathbf{z} = [\theta, \tau, v, dq]^{\mathrm{T}}$, while Case C extends the state vector with the parameter ζ, i.e., $\mathbf{z} = [\theta, \tau, v, dq, \zeta]^{\mathrm{T}}$. The measurement vector has two elements and contains the (synthetic) measurements of fault shear stress and slip rate.

We sample synthetic measurements from the truth run with parameter $\zeta_t = 0.7$. We assume the state variables of shear stress (τ) and slip-velocity (v) to be measured every 4 time units. Measurement errors are sampled from a normal distribution and added to the truth-run samples. For shear stress, the standard deviation of this normal distribution is 0.6. The standard deviation of the measurement errors added to the slip velocity measurements is 1.15.

We generate the prior particles with a biased $\zeta = 0.6$. In the following, we denote this biased parameter with ζ' and the true parameter as $\hat{\zeta}$. We evaluate the data-assimilation performance by comparing the posterior of $\mathbf{z}$ to the values of $\mathbf{z}$ in the true trajectory. Figure 19.2 illustrates the state trajectories for $\hat{\zeta} = 0.7$ and $\zeta' = 0.6$.

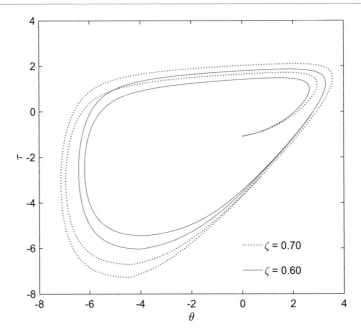

Fig. 19.2 Phase diagram for a Burridge–Knopoff (BK) model with rate-and-state-friction law when $\zeta = 0.6$ (bias case) and $\zeta = 0.7$, i.e., a case of no bias

Starting from an initial condition of $\theta = 0$ and $\tau = -1$, both θ and τ increase, after which the shear stress τ stays constant and θ drops (the interseismic phase). Following this phase, the shear stress drops while θ remains stable in the coseismic phase. After that, we observe a very short post-seismic phase where τ remains low and θ rapidly becomes less negative. The system then moves to the interseismic phase in which both θ and τ increase again and the cycle repeats. For both $\zeta = 0.6$ and $\zeta = 0.7$, the trajectories are stable cycles. Figure 19.2 shows that the trajectory for $\zeta = 0.6$ has smaller amplitudes of τ and θ than the trajectory for $\zeta = 0.7$. A smaller value of ζ results in a shorter cycle.

The experiments use $N = 1000$ particles in a standard particle filter with importance resampling similar as described in Sect. 9.2.1. We assume the likelihood $f(\mathbf{d}|\mathbf{z}_j)$ in Eq. 9.4 to be known (see Algorithm 9). We use a Lorentz function rather than a Gaussian (see Algorithm 9), i.e., for a measurement d and the model state $\mathbf{z}$ in assimilation window l, the likelihood becomes

$$f\left(d_l|\mathbf{z}_l\right) = \frac{1}{1 + \dfrac{\left(d_l - H(\mathbf{z}_l)\right)^2}{\sigma_l^2}}, \tag{19.4}$$

with the variance σ_l^2 of the measurement noise. This function has the advantage of broader tails than the Gaussian function, resulting in fewer particles having zero weight (Vossepoel & Van Leeuwen, 2007). While this choice helps the particle filter algorithm avoid degeneracy, it is slightly inconsistent with the actual measurement errors drawn from a Gaussian with the same width.

19.4   Case A: State Estimation

Figure 19.3 illustrates the evolution of the posterior mean τ, θ and v, in the presence of a bias ($\zeta' = 0.6$) compared to the truth case of no bias ($\hat{\zeta} = 0.7$). The figure illustrates that when performing only state estimation while not accounting for the parameter bias, the data assimilation cannot fully recover the true trajectory of these state variables. While the data assimilation appears to decrease the frequency of the events and brings the shear stress and θ closer to the truth, the resulting analysis does not capture the amplitude of the τ and θ cycles very well, and the reconstruction of the slip rate v is poor.

During the interseismic phase (e.g., at $t = 16$), the difference between prior and truth is not very large. But at time $t = 20$, i.e., in the coseismic phase, the distribution of the prior shear stress has a peak around $\tau = -5$, while the weights

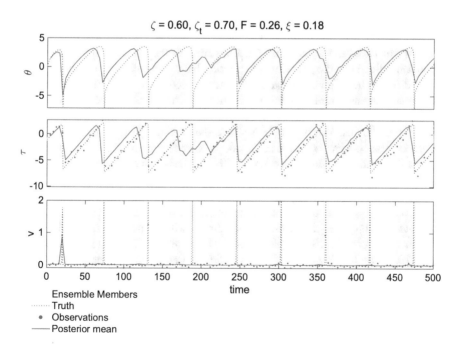

Fig. 19.3 Case A: Evolution of 1000 particles and the posterior mean and truth of state variables θ, shear stress τ, and slip velocity v in an experiment with state estimation for a synthetic experiment where the prior has a biased parameter relative to the truth. The dotted blue lines provide the actual θ, τ, and v, and cyan dots represent synthetic measurements. Grey lines represent the prior particles, and the red line is the analysis given by the weighted mean of the posterior distribution

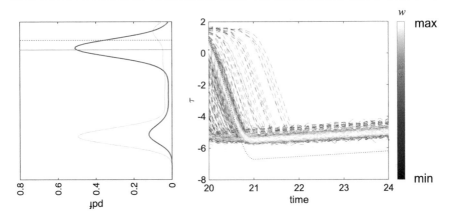

Fig. 19.4 Time evolution of 1000 particles (right panel), as shown in Fig. 19.3, zoomed in for $t = 20$. The colors of the lines represent the weights for $t = 20$. The lines have a lighter color when the particle they represent has a higher weight. The neighboring left panel graph gives the distribution of prior particles (in grey) and the distribution after resampling (in red)

at that moment in time are highest for the shear stress value around $\tau = 0$ (see Fig. 19.4). The weighting of the particles favors the particles whose shear stress has not yet dropped. The relatively late change in shear stress as measured only occurs in a few particles, and as a result, the filter reaches the criterion for resampling. Figure 19.4 illustrates the particles' distribution before and after resampling.

The different behavior of the particle filter compared to the ensemble Kalman filter (EnKF) is illustrated in Fig. 19.5. For the times $t = 20$, $t = 24$, $t = 72$, and $t = 76$, this figure presents the pdfs for Case A with the standard particle filter and the EnKF. It shows that the posterior distribution of τ in the case of a particle filter can feature small secondary peaks next to the dominant peak, and it can redistribute the particles in such a way that sudden transitions in the trajectory, such as in the coseismic phase at $t = 20$, are captured in the posterior. In the EnKF experiment, the posterior distribution of τ remains very close to its prior distribution. For the interseismic period, the results of the particle filter and the EnKF are virtually the same.

Figure 19.6 illustrates the particle results further with a phase diagram of the θ and τ values of the particles. These plots give the expected phase trajectories, as shown in Fig. 19.2. The lower value of ζ' in the prior particles implies a phase difference compared to the truth. We can only partly compensate for this difference by favoring particles that quickly enter the coseismic phase. We also observe that for $t = 20$, the particles closest to the truth in the τ-θ space obtain the highest weight. At $t = 24$, $t = 72$, and $t = 76$, this is less so because of the match of the slip rate v to the truth (not shown). By adjusting the state, the assimilation partly compensates for the bias in the parameter, but the resulting posterior fails to follow the actual trajectory.

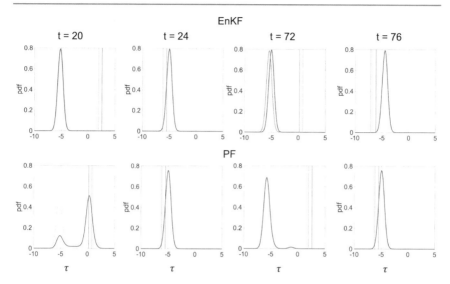

Fig. 19.5 Case A: Comparison of EnKF (top) and particle filter (bottom) for state estimation in the presence of a biased prior parameter for $t = 20$, $t = 24$, $t = 72$ and $t = 76$. The panels give the prior distribution (grey for the particle filter, blue for EnKF) and posterior distribution (red) of τ in the ensemble members/particles. The value of the measured shear stress is indicated with a pink line, the true shear stress with a dotted black line

19.5 Case B: State Estimation with Increased Model Error

In many real-life applications, it is not clear beforehand whether the difference between model and measurements comes from errors in the model state, in the model parameters, the model controls, or in the model itself. In these situations, one typically applies the model as a weak constraint. In Case B, we investigate whether we can compensate for a bias in the prior parameter by increasing the prior model error in Eq. (19.2) from 0.01 to 0.1.

Figure 19.7 shows the results. Not surprisingly, a more significant model error increases the spread in the prior compared to Case A (Fig. 19.6). The most noticeable difference occurs at $t = 72$, just before the coseismic phase. Some particles still have high shear-stress values, while the shear stress has dropped for others. By giving the high-shear-stress particles a relatively high weight, the particle filter with this weak-constraint formulation partly compensates for the bias in the parameter.

19.6 Case C: State- and Parameter Estimation

In Case C, we investigate state- and parameter estimation in the case of a biased prior parameter. The state vector contains the state variables τ, θ, and v and the parameter ζ. We generate the initial ensemble by sampling each particle from a log-

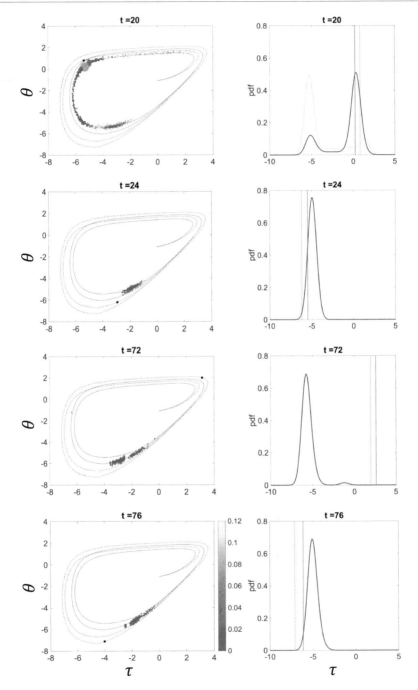

Fig. 19.6 Case A: Left column: phase diagrams showing the θ-τ trajectories of the true state (black, $\hat{\zeta} = 0.7$) and the biased prior (blue, $\zeta' = 0.6$) as in Fig. 19.2. Particles of the Case A experiment are indicated with dots, whose size and color indicate their weights for $t = 20, 24, 72,$ and 76. We indicate the true values of θ and τ for these times with a black dot. Right column: the posterior (red) and prior (grey) shear stress distributions. For $t = 24, 72,$ and 76, the difference between the prior and the posterior is so small that the two lines overlap

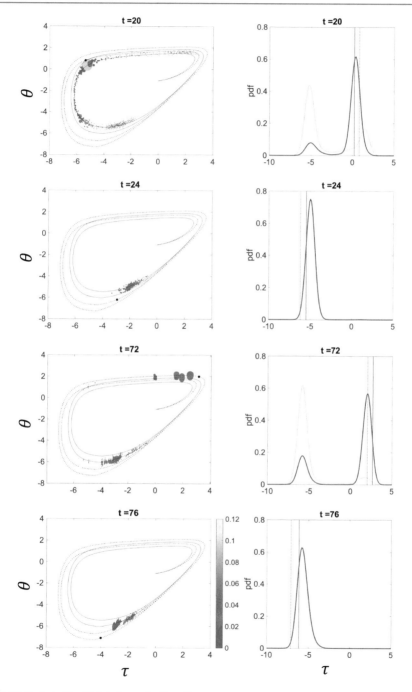

Fig. 19.7 As in Fig. 19.6, but now for Case B

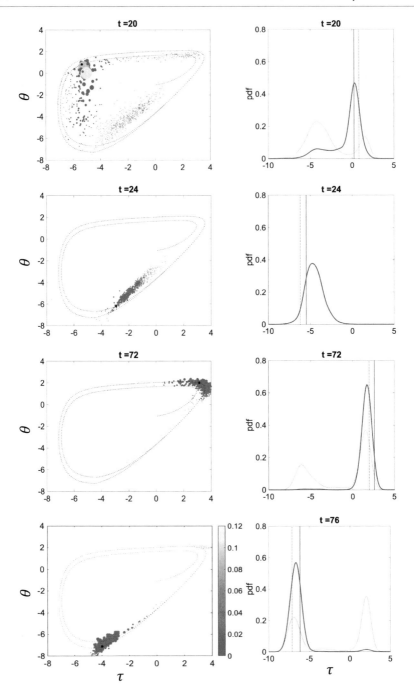

Fig. 19.8 As in Fig. 19.6, but now for Case C

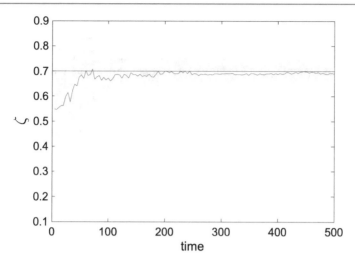

Fig. 19.9 Case C: evolution of the parameter estimation

normal distribution with mean ζ' and standard deviation 0.01. To ensure that the state of each particle is different but consistent with its parameter value, we sample its state from a single realization with this parameter at a random moment in time. This approach is similar to the lagged forecasting by Hoffman and Kalnay (1983). The resulting N particles have N parameter values centered around ζ' and different initial states that are consistent with the parameter of each particle.

Figure 19.8 provides the phase diagrams for Case C's state- and parameter estimation. The distributions in this figure demonstrate that the state- and parameter estimation results in a posterior closer to the truth than in Case A or B. By varying parameter values within the ensemble, the resulting states and phases within the seismic cycle cover a more significant part of the θ-τ domain. The results suggest that, for this case, state- and parameter estimation provides an effective reconstruction of the true τ and θ. However, for the time steps $t = 24, 72$, and 76, none of the particles represented v accurately, and as a result, the weights remained relatively low. Figure 19.9 shows the reconstruction of ζ. After a period with significant adjustment of the posterior parameter value, the estimation of the parameter fluctuates around the true value, and the true value is within the ensemble spread. For this particular case, there is no degeneracy of the ensemble. An artificial evolution equation for the parameters could have resolved the problem if we observed degeneracy.

19.7 Summary

These results indicate that for the particular case of seismic-cycle estimation presented in this chapter, the combined state- and parameter estimation provides a more favorable data-assimilation outcome in the presence of a parameter bias than state estimation alone. Increasing model error in state estimation can only partly compensate for this bias. We also see that a particle filter handles bimodal pdfs and provides physically acceptable solutions while the EnKF fails. The low dimension of the current example makes the use of a standard particle filter straightforward.

Particle Flow for a Quasi-Geostrophic Model

20

This chapter discusses the application of a particle-flow filter to a two-layer quasi-geostrophic model. The reasons for including this example are twofold. First, it shows that it is possible to apply fully nonlinear data assimilation to high-dimensional systems, even without localization. Second, we introduce how to set up such an experiment in more detail and discuss the choices one must make.

20.1 Introduction

We start by demonstrating how the evolution of the particles in pseudo-time progresses in a highly idealized model to understand the basic ideas behind particle-flow methods better. We look at one specific gridpoint in a 1000-dimensional Lorenz 1996 model for an observation that is the square of the state variable at that gridpoint, so $d = x_{\text{true}}^2 + \epsilon$. The value of $d = 7.3$ with observation-error standard deviation equal to 0.2. The prior is a wide Gaussian with mean 0.5 and standard deviation 1, represented by 100 particles as depicted by the lower red dots in Fig. 20.1. The blue lines denote the movement of the particles in the one-dimensional state space of this gridpoint. The vertical axis is pseudo time, scaled between 0 and 1. We made many iterations with small steps to accurately illustrate the movement in this part of the state space.

Figure 20.1 shows that the particles flow towards the posterior pdf, centered on the possible gridpoint values corresponding to $x^2 = d = 7.3$, hence $x = \pm\sqrt{7.3} = \pm 2.7$, with a standard deviation of order 0.1. The pseudo-time trajectories seem to cross each other, e.g., in the lower right corner of the plot. An actual crossing of trajectories would lead to the failure of the method, indicating the use of too large pseudo time steps. However, this crossing is not actual because we only plot

© The Author(s) 2022
G. Evensen et al., *Data Assimilation Fundamentals*,
Springer Textbooks in Earth Sciences, Geography and Environment,
https://doi.org/10.1007/978-3-030-96709-3_20

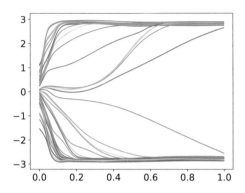

Fig. 20.1 The plot shows the evolution of 100 particles (colored lines) in pseudo time (horizontal axis) at one gridpoint in a 1000-dimensional Lorenz 1996 data-assimilation experiment. Note the motion from the relatively narrow prior distribution to the two modes

the flow of the 1000-dimensional particles in a one-dimensional projection. In the 1000-dimensional full space, the particles do not cross.

The particle-flow filter demonstrates behavior that is impossible to obtain with an ensemble Kalman filter, not even an iterative ensemble Kalman filter that uses the ensemble gradient for the adjoint of the observation operator. The reason is that the gradients will have different signs for the two posterior modes, while the ensemble provides only one average gradient. Only iterative methods that use either the adjoint of H or different ensemble gradients for different state-variable values can accurately find the two modes. Variational methods will converge to one of the modes, dependent on the first guess. Finally, a standard particle filter will not move the particles, only their weights. The relatively narrow prior in Fig. 20.1 does not cover the two posterior modes, and no particles will end up in these modes. Resampling would produce two artificial modes at the extremes of the prior. Only a particle-flow filter can produce these modes in its standard configuration.

20.2 Application to the QG Model

The quasi geostrophic (QG) model solves the following equations for a 2-layer system

$$\frac{\partial p_1}{\partial t} + J(\psi_1, p_1) = A \Delta q_1, \tag{20.1}$$

$$\frac{\partial p_2}{\partial t} + J(\psi_2, p_2) = A \Delta q_2, \tag{20.2}$$

where the potential vorticity p_i in each layer is the sum of the relative vorticity, the planetary vorticity, and a stretching term,

$$p_1 = \nabla^2 \psi_1 + f - F_1(\psi_1 - \psi_2), \tag{20.3}$$

$$p_2 = \nabla^2 \psi_2 + f + F_2(\psi_1 - \psi_2). \tag{20.4}$$

Here ψ_1 and ψ_2 are the stream functions in the two model layers, and $\mathbf{A}$ is the horizontal diffusion or mixing coefficient. The Jacobian $J(\psi, p) = \frac{\partial \psi}{\partial x}\frac{\partial p}{\partial y} - \frac{\partial \psi}{\partial y}\frac{\partial p}{\partial x}$

denotes the advection of potential vorticity. The Coriolis parameter is $f = f_0 + \beta y$ in which y is the meridional coordinate (the so-called β-plane approximation), and the F_i are constants related to the densities and height of the two layers.

A practical scheme to solve the QG model equations is the following. First, calculate the potential vorticity from the stream-function fields. Next, propagate the potential-vorticity fields over one time step. Then solve the Helmholtz equations for the new stream-function fields, which the advection terms use to propagate the potential vorticity over the next time step.

The model setup uses two layers of 257 by 129 gridpoints with a grid spacing of 100 km. The dimension of the state vector is 66306. The time step is 30 min, and $F_1 = F_2 = 2.8 \times 10^{-12}$ m^{-2}. The Coriolis parameter in the middle of the domain is $f_0 = 7.28 \times 10^{-5}$ s^{-1} and $\beta = 2.0 \times 10^{-11}$ m^{-1}s^{-1}.

20.3   Data-Assimilation Experiment

We initialized the model with a meandering jet of wavenumber four in the upper layer with maximum stream-function value 5×10^7 m^{2}s^{-1}, and the stream function in the lower layer was taken as a factor 0.03 times that of the upper layer. This model state was spun up for 250 time steps, approximately five days.

Figure 20.2 gives examples of the true model stream function in the upper layer at different time steps during the data-assimilation experiment. The plots show different stages of the evolution of the flow field, with the Jet Stream flowing from East to West at the boundary between reddish and greenish colors. We observe several eddies (low- and high-pressure cells) north and south of the meandering Jet Stream. The three plots show the shedding of high-pressure cells for a little more than two days.

An initial ensemble of 100 members was created by adding Gaussian random noise with a decorrelation length scale of 20 gridpoints to a similarly perturbed true spun-up state. The standard deviation of the perturbations was 100 m^{2}s^{-1}. At every time step, we added model errors drawn from a Gaussian with zero mean and the same decorrelation length scale and a standard deviation of 0.005 times the one in the true initial fields.

We assimilated observations every 10 time steps, corresponding to 5 h. We observed the stream function at 600 equally-distributed gridpoints in each layer. This number corresponds to a fraction of 0.036 of the total number of gridpoints. Observation errors were uncorrelated, with standard deviation 5×10^5 m^{2}s^{-1}.

To understand what else is needed, we show the evolution equation for the particles in pseudo time s,

$$\frac{d\mathbf{x}_j}{ds} = \mathbf{D}\frac{1}{N}\sum_{l=1}^{N}\left(\mathbf{K}(\mathbf{x}_j, \mathbf{x}_l)\nabla_\mathbf{x}\log f(\mathbf{x}_l|\mathbf{d}) + \nabla_{\mathbf{x}_l}\mathbf{K}(\mathbf{x}_j, \mathbf{x}_l)\right). \tag{20.5}$$

This equation shows we need to provide three ingredients: the likelihood $f(\mathbf{d}|\mathbf{x})$, a continuous version of the prior $f(\mathbf{x})$, and the matrix-valued kernel $\mathbf{K}$. We assume the likelihood is known, and we do have a representation of the prior by a set of particles.

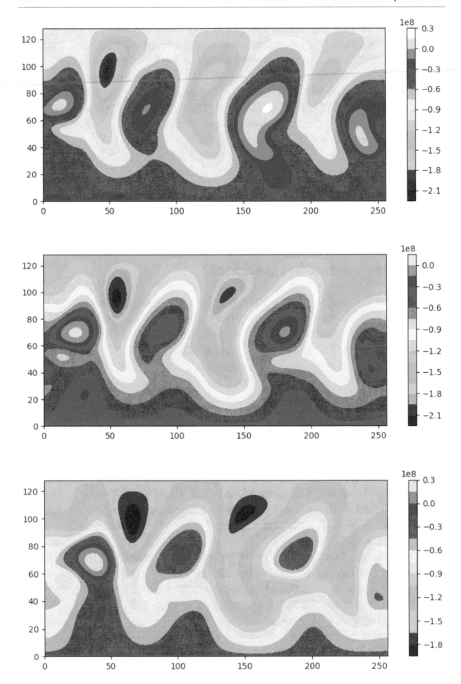

Fig. 20.2 The plots show the upper-layer stream-function fields from the true run 25 h apart from top to bottom

In the evolution equation for the particles, we need to take the gradient of the prior pdf to the state **x**, so a representation in terms of delta functions is not sufficient. Several possibilities for approximations are possible. One is to assume the prior is a Gaussian mixture model centered on the particle positions. Another is to use a single Gaussian as the prior pdf. Note that this approximation is only needed to find an approximate gradient of the prior. The prior particles can still represent a non-Gaussian pdf. This situation is similar to EnKF, which updates each ensemble member separately. In the EnKF, the posterior pdf can retain non-Gaussian structures present in the prior ensemble even though it uses a Gaussian approximation to define the update. In this application, we assumed that the prior particles represent a Gaussian pdf, defined by the ensemble mean and ensemble covariance.

We use a matrix-valued kernel with off-diagonal entries equal to zero, and on the diagonal a scalar Gaussian kernel,

$$k_{ii}\left(\mathbf{x}_j, \mathbf{x}_l\right) = \exp\left(-\frac{1}{2}\frac{\left(x_j^i - x_l^i\right)^2}{\sigma_i^2}\right),\tag{20.6}$$

where σ_i^2 is the prior variance in state variable i. In the limit of an infinite number of ensemble members, theory tells us that any smooth, symmetric kernel will result in the prior particles converging to the posterior pdf. In practice, with a small number of particles, care has to be taken to ensure fast convergence.

20.4   Results

Figure 20.3 shows the prior mean, the truth, and the posterior mean of the lower layer stream-function fields at day 10 of the assimilation experiment as an example of the outputs. The posterior mean is indeed much closer to the truth than the prior mean, as expected. The data assimilation manages to deepen low-pressure areas, make high-pressure regions less deep, and generate a more accurate splitting of the Jet Stream around the gridpoint (200, 70).

Figure 20.4 compares the time evolution of the spatially averaged mean-square errors of the ensemble mean and the ensemble variance. We see the typical decrease of errors at assimilation times and the growth of the actual and predicted errors between assimilation times. The two curves closely follow each other, showing that the ensemble spread is a realistic estimate of the true error (defined as the square of the difference between ensemble mean and the truth run).

The particle-flow filter is an iterative scheme that reduces the KL-divergence at every time step. To illustrate this property, the right plot in Fig. 20.4 shows the mean square error spatially averaged between the ensemble mean and the truth as a function of the iteration number. Each line corresponds to a different observation time. The error converges to a fixed value, mainly determined by the observation error. For practical reasons, we limited the number of iterations to 50, but additional iterations could have reduced the divergence even further.

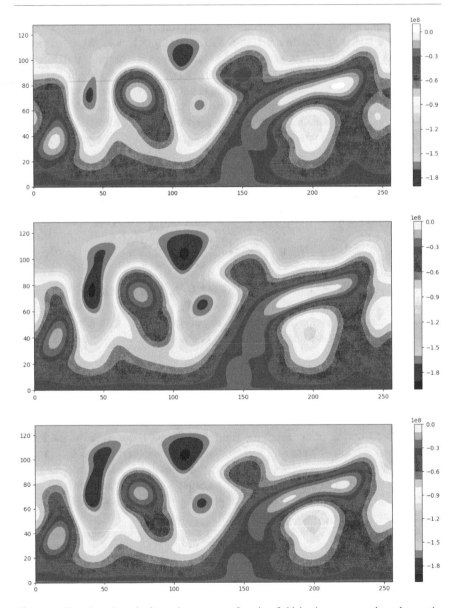

Fig. 20.3 The plots show the lower-layer stream-function fields' prior mean, truth, and posterior mean at day ten from top to bottom

We obtained these experimental results by observing the stream-function value directly at 600 points in each layer at each observation time. It is interesting to see what happens when using a nonlinear observation operator. We also performed an experiment where we observed the square of the stream function at each observation point. This situation typically leads to a skewed posterior pdf when all prior particles

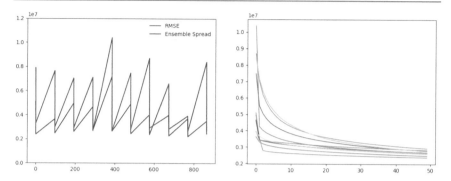

Fig. 20.4 The left plot shows the time evolution of the true error (red line) and the ensemble variance (blue) over 19 d. The right plot shows the convergence of the ensemble mean to the truth as a function of the pseudo time step (or iteration). The different curves correspond to the observation times in the left plot

are positive. The likelihood is bimodal, but the prior only sees one of the modes. The skewness arises because of the nonlinear transformation between state and observation space. The more exciting situation appears in observed gridpoints where prior particles have different signs for the stream-function value. In that case, both the positive and the negative root of the observation are covered by the prior. Thus, the likelihood will be bimodal in the domain where the prior is non-zero. Figure 20.5 depicts what can happen in such a case. Since the observation is the square of the stream-function value, it points to two possible solutions, one positive and one negative. The blue histogram represents the prior pdf. The red bars indicate the possible values of the observation at this gridpoint, and the orange histogram represents the

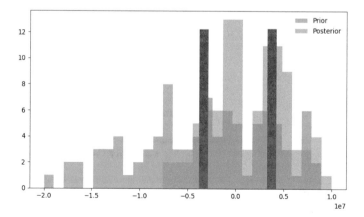

Fig. 20.5 The figure shows the QG model's particle-flow filter results in a selected gridpoint using a quadratic observation operator. The two red bars denote the two possible positions of the observation in state space. The prior (blue) and posterior (orange) histograms represent the distribution of the 100 prior and posterior particles

posterior pdf. The prior pdf is a wide pdf with no particular structure. The likelihood in state space is bimodal, and the posterior pdf is indeed bimodal as expected.

This example demonstrates how to set up a particle-flow filter in a large-dimensional system. Localization is not needed explicitly. Research on these methods is still in its infancy, but fully nonlinear data assimilation seems to have come within reach.

EnRML for History Matching Petroleum Models

21

In this chapter, we present an application of an iterative ensemble smoother for a history-matching case with a reservoir simulator. The application is realistic and represents an actual oil reservoir with production data. This case focuses on formulating the history-matching problem with consistent error statistics. The chapter shows how we can use ensemble methods to estimate high-dimensional parameter sets and additional model controls by conditioning the model on fluid production rates.

21.1 Reservoir Modeling

In petroleum engineering, reservoir engineers use parameter-estimation methods to improve the characterization of oil reservoirs. Oil reservoirs are permeable layers in the subsurface that are bounded by structural or stratigraphic elements that create seals (traps) on the reservoir's top and sides. For example, the seals can be, e.g., shale layers with low permeability or impermeable faults. We have only limited knowledge of the reservoir properties. We obtain coarse information about the large-scale reservoir structure from seismic data, and we have localized point information from core samples from test wells. With additional assumptions about the depositional environment, it is possible to build a geologic model of the reservoir. However, the model will always be an approximation of reality.

The geologic model and the seismic data form the basis for well planning, drilling and production. Therefore, any improvement of the reservoir model can significantly impact the reservoir economy. For example, one can define a reservoir-simulation model used to simulate the production of existing and planned wells from the geologic model. Typically, the predicted oil production is initially vastly different from the actual oil production due to flaws in the reservoir model.

© The Author(s) 2022
G. Evensen et al., *Data Assimilation Fundamentals*,
Springer Textbooks in Earth Sciences, Geography and Environment,
https://doi.org/10.1007/978-3-030-96709-3_21

21.2 History Matching Reservoir Models

The reservoir model has the form of Eq. (2.34), which we write as

$$\mathbf{y} = \mathbf{g}(\mathbf{z}) = \mathbf{h}\big(\mathbf{m}(\mathbf{x}, \boldsymbol{\theta}, \mathbf{u})\big) \tag{21.1}$$

Here $\mathbf{x}$ denotes the initialization of the dynamical variables (e.g., oil, water, gas, pressure), $\boldsymbol{\theta}$ represents all the uncertain reservoir parameters such as the three-dimensional porosity and permeability fields, fault multipliers, structural surfaces, etc. The uncertain control variables in $\mathbf{u}$ represent the production of oil, water, and gas from the production wells and the injection of water and gas through the injection wells. For predictions, we specify the controls in $\mathbf{u}$, while for a historical simulation, we use the observed well-rates in $\mathbf{u}$. Typically, one assumes that the uncertain model parameters in $\boldsymbol{\theta}$ dominate the model errors, and we have set $\mathbf{q} = 0$. In Eq. (21.1), $\mathbf{y}$ represents the predicted measurements corresponding to each well's produced oil, water, and gas.

We define the data-assimilation problem as estimating or updating the uncertain model initial conditions $\mathbf{x}$, model parameters $\boldsymbol{\theta}$, and model control parameters $\mathbf{u}$, given the prior information and the observed production and injection. If the reservoir could deliver all the production data enforced through $\mathbf{u}$, there would be no misfit between observed and predicted data and consequently no model update. However, the reservoir model can generally not deliver the observed historical production data enforced on the model. Thus, we update model parameters to fit the observed production better. Hence, the name *history matching*.

Methods for parameter estimation in petroleum engineering typically sample the posterior pdf in Eq. (2.43) while assuming Gaussian priors and neglecting the model errors $\mathbf{q}$. For this joint parameter-state estimation problem, a filtering approach requires recursive updates of the parameters and dynamical state, which typically introduces dynamical inconsistencies and adds to the computation time by numerous stops and restarts of the model (Evensen et al., 2007; Gu & Oliver, 2005; Haugen et al., 2008; Reynolds et al., 2006; Seiler et al., 2007; Skjervheim et al., 2009). For this reason, Skjervheim et al. (2011) introduced the use of ensemble smoothers for reservoir history matching. Following Skjervheim et al. (2011), there was a rapid development of iterative ensemble smoothers such as the EnRML (Chen & Oliver, 2012, 2013) and the ESMDA (Emerick & Reynolds, 2013). Recent papers (Evensen, 2018, 2019, 2021; Evensen & Eikrem, 2018; Evensen et al., 2019; Raanes et al., 2019) have analyzed and further developed the iterative smoothers and enhanced their performance.

21.3 Example

We will now present an example from Evensen (2021), who discussed the consistent formulation of the history-matching problem and illustrated its solution. Evensen (2021) particularly emphasized that one needs to consider the uncertainties of the model controls and include their temporal error correlations to compute a consistent update. The model was a realistic but straightforward reservoir model with six producing wells and three injectors. The uncertain parameters included the model porosity and seven fault multipliers. Evensen (2021) found that by updating the porosity field, the fault multipliers, and the model controls, one obtained an updated ensemble of models that fit the production data within their prescribed uncertainty. The assimilation method was the subspace EnRML from Algorithm 5.

Figures 21.1 and 21.2 present some history-matching results from a case where we have assumed substantial time correlations in the rate errors, which Evensen (2021) found to be the most realistic. He sampled the prior fault-multiplier realizations from a log-uniform distribution on the interval 0.001 to 1.0 for all the faults. Thus, with a log-scale on the y-axis, the samples would appear uniformly distributed. In Fig. 21.1, the circles of different colors denote the updated ensembles of multipliers. Notably, the F3, F4, and F5 faults are closed after the conditioning. Evensen (2021) also updated the model's three-dimensional porosity field (not shown). The left plots in Fig. 21.2 show the prior and posterior ensembles of model-predicted oil, gas, and water production rates from top to bottom. The plots to the right in Fig. 21.2 present the prior and posterior historical rates, which we use as control variables in the simulation model. We observe a weak reduction in the ensemble variance for OPR and WPR. At the same time, for GPR, there is a significant update with both reduced gas production and a lower posterior ensemble variance. The updated model parameters and controls result in the posterior ensemble prediction shown by the red curves. The posterior ensemble fits the observations within their two standard deviations error bars.

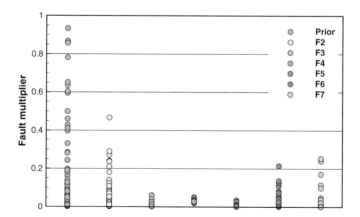

Fig. 21.1 Petroleum case: Prior and posterior fault multiplier realizations. The prior distribution is log-uniform on the interval 0.001 to 1.0 for all faults

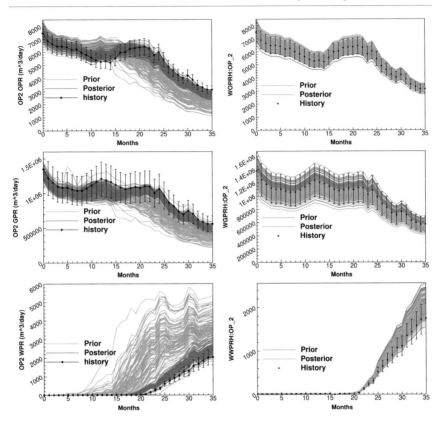

Fig. 21.2 Petroleum case: The plots show the prior and posterior ensembles of predicted and observed production of oil (OPR), gas (GPR), and water (WPR) for the well OP2, from top to bottom. The left plots show the ensemble of predicted rates, while in the right column we present the ensemble of historical rates used to force the model. The green curves are the prior realizations of predicted rates, and the red curves are the corresponding updated realizations. The blue curves are the prior control-rates realizations used to force the model, while the orange curves are the updated realizations

We point out that the example from Evensen (2021) is the first time the conditioning process includes the model controls as variables to be updated. This approach resolves previously reported issues related to overfitting the measured rates and underestimating the posterior ensemble variance. We refer to the paper for a detailed discussion.

ESMDA with a SARS-COV-2 Pandemic Model

22

In parallel with the global SARS-COV-2 pandemic, several data-assimilation practitioners have implemented systems to predict pandemic development. Common for several of these studies is an approach of recursive updating of the model's state variables using EnKFs or even particle filters. However, the problem is closer to a parameter-estimation problem than state estimation. Evensen et al. (2020) presented a new assimilation process for predicting the SARS-COV-2 pandemic. They used an approach of combined parameter and control-variable estimation. The reason is that the effective reproductive number $R(t)$ drives the model. And it is a function of time, as it is the public behavior that determines its value. Another exciting aspect is that today's number of observed hospitalizations and deaths results from people's behavior about two weeks earlier. In their paper, Evensen et al. (2020) showed that the assimilation system, which used ESMDA to estimate parameters and the time-dependent control, $R(t)$, was capable of tracking the pandemic over multiple "waves" and making predictions with   realistic uncertainty estimates.

22.1 An Extended SEIR Model

Evensen et al. (2020) developed an extended SEIR (susceptible, exposed, infectious, and recovered) model (Blackwood & Childs, 2018) for predicting the SARS-COV-2 pandemic. The model has multiple age classes (since the COVID-19 disease affects different age groups differently) and includes compartments for quarantined, hospitalized, and dead, with additional separation into those with mild, severe, and fatal symptoms.

Figure 22.1 gives an overview of the model where we have stratified the susceptible, exposed, and infectious populations into age groups $\mathbf{S}_i$, $\mathbf{E}_i$, and $\mathbf{I}_i$ following Cao and Zhou (2012). As in the standard SEIR model, the infectious and susceptible

© The Author(s) 2022
G. Evensen et al., *Data Assimilation Fundamentals*,
Springer Textbooks in Earth Sciences, Geography and Environment,
https://doi.org/10.1007/978-3-030-96709-3_22

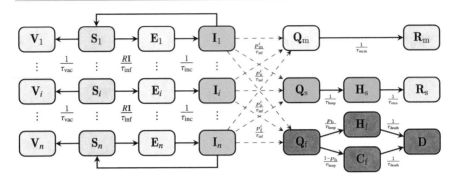

Fig. 22.1 Flow diagram of the SEIR model

interaction leads to the newly exposed. The effective reproductive numbers between different age groups R_{ij} together with the infection time scale τ_{inf} determine the rate of new infections. We will discuss the formulation used for R_{ij} in detail below. Note that the susceptible and infectious interaction constitutes the only source of nonlinearity in the model.

The different age groups of infectious $\mathbf{I}_i$, transition into the various groups of quarantined sick, $\mathbf{Q}_{\text{m}}, \mathbf{Q}_{\text{s}},$ and $\mathbf{Q}_{\text{f}}$, based on the fractions $p_{\text{m}}^i, p_{\text{s}}^i, p_{\text{f}}^i$, and the infection time scale τ_{inf}. The fractions refer to patients with mild symptoms, hospitalized patients with severe symptoms, and fatally ill patients and specify how the virus affects people in different age groups. The subscripts m, s, and f refer to mild, severe, and fatal symptoms. Thus, the model includes different probabilities for dying or being hospitalized dependent on the age group and accounts for how the SARS-CoV-2 virus affects older people more severely. We have assumed that a patient will not infect anyone while in a quarantined group.

The patients with mild symptoms in $\mathbf{Q}_{\text{m}}$ will recover and transition into the group of recovered with mild symptoms $\mathbf{R}_{\text{m}}$, on a time scale τ_{recm}, without going to the hospital. Severely sick patients in $\mathbf{Q}_{\text{s}}$ transfer to the hospital compartment $\mathbf{H}_{\text{s}}$, on a time scale τ_{hosp}. After that, they recover on a time scale τ_{recs} into the compartment of patients recovered from severe disease $\mathbf{R}_{\text{s}}$.

The model admits the fatally-ill patients in $\mathbf{Q}_{\text{f}}$ to a hospital $\mathbf{H}_{\text{f}}$ on the time scale τ_{hosp}. However, we also allow for a fraction of fatally-ill patients not admitted to a hospital, and we model them in $\mathbf{C}_{\text{f}}$. The purpose of the $\mathbf{C}_{\text{f}}$ variable is to include the fatally-ill patients not measured as hospitalized. Introducing $\mathbf{C}_{\text{f}}$ allows us to use realistic fractions p_{f} of fatally-ill patients and still condition on the measured hospitalization numbers $\mathbf{H}_{\text{s}} + \mathbf{H}_{\text{f}}$. This partition of the fatally-ill patients was essential for most cases discussed in the paper. The fatally ill patients in $\mathbf{H}_{\text{f}}$ and $\mathbf{C}_{\text{f}}$ end up in the group of dead $\mathbf{D}$ on a time scale τ_{death}. Later, we added compartments of vaccinated $\mathbf{V}_i$, which was essential to match the data after the vaccinations started in 2021.

A challenging property of the model is that the time-dependent effective reproductive number $R(t)$ is the primary driver of its evolution. Furthermore, $R(t)$'s value about two weeks ago determines today's number of hospitalizations and deaths. Hence, standard sequential state estimation, Sect. 2.4.2, is not appropriate for this problem, as the model will be strongly biased unless we correct $R(t)$. It turns out that the best approach to solve this problem is to consider it as a combined initial-value-, parameter-, and model-control estimation problem. This application corresponds to the problem definition in Sect. 2.4.6. Thus, we define the assimilation window to include the simulation period from the pandemic's onset until today. We estimate the initial number of exposed and infectious, i.e., the initial conditions, and all the model's static time scales. Still, the dominating parameter in the model is the control variable $R(t)$. In the model example below, we also included the impact of vaccinations. Unless we explicitly model the vaccination, it is impossible to fit the model prediction to the observations.

22.2 Example

In this application, we used ESMDA with $M = 32$ steps and a vast ensemble size of $N = 5000$ realizations (see the sensitivity experiments below). In Fig. 22.2, we show results from one such experiment for Norway. The upper plot shows the ensemble predictions of current hospitalizations in red and accumulated deaths in green, and the black dots are the observations. The ensemble of effective $R(t)$ controls, shown in blue, drives the time evolution of the pandemic. We note the immediate sharp drop in $R(t)$ in mid-March when Norway shut down, and we estimate $R(t)$ values around 0.4 at the end of March. This reduction of $R(t)$ to below one leads to an efficient decline in hospitalizations and deaths. The pandemic was essentially over at the end of July. Then, during July and August, the government allowed for vacation travels outside Norway, and during August and September, many foreign workers came back to Norway. Norway had, at this time, no effective quarantine system, and the virus

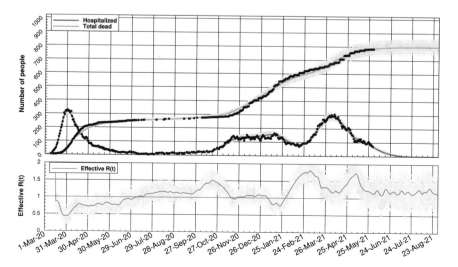

Fig. 22.2 Ensemble predictions for Norway including the impact of vaccinations

started spreading again throughout society with several prominent local outbreaks. We see that the estimated $R(t)$ is above one from the second half of July until the end of the year. This increase in $R(t)$ accounts for the relatively weak restrictions on social contact between people and the additional imported cases as we do not explicitly model them. The increase in $R(t)$ led to the second wave during November and December, followed by a strict lockdown in January. This new lockdown initially seemed to control the pandemic and reduced the number of cases. However, during March 2021, we have experienced another steep increase in infections caused by higher values of $R(t)$ partly caused by the introduction of mutated, more infectious viruses. The vaccinations of the adult population, which started in January, helped control the further pandemic growth.

22.3   Sensitivity to Ensemble Size

We have examined the convergence of ESMDA as a function of the ensemble size. ESMDA being a Monte Carlo algorithm means that we can continually improve

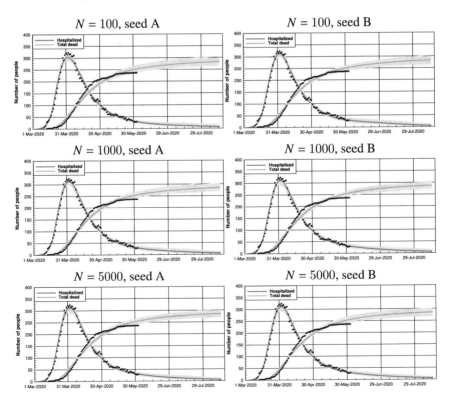

Fig. 22.3 The plots show the estimated solution of current hospitalized in red and accumulated deaths in green for increasing ensemble sizes of $N = 100$, $N = 1000$, and $N = 5000$. The black dots are the measurements. The left and right columns show results from two random seeds. All the cases used 32 MDA steps

the solution by increasing the ensemble size. However, we need to decide on a tradeoff between ensemble size and the number of ESMDA steps due to limitations on computing power. While the number of ESMDA steps impacts the actual convergence of the algorithm towards the right solution, the ensemble size impacts the precision of the statistical estimate of the final solution. We find it most important to first converge to the correct physical solution and, then, use an as large as possible ensemble to reduce the sampling errors.

Figures 22.3 and 22.4 show the results using various ensemble sizes and two different random seeds. For all cases, we used a sufficient number $N_{MDA} = 32$ to ensure a converged ESMDA solution. These plots demonstrate the robustness of ensemble-based assimilation methods. Even using 100 realizations, the posterior predictions are consistent with the data and very similar to the cases with larger ensemble sizes. There is a visual effect on the update from the random seed that we see more clearly in the estimated $R(t)$. When using 1000 realizations, there is still a significant difference in the estimated R(t). It is hard to note any dependency on the seed or difference from the case with 5000 realizations for the predictions. When

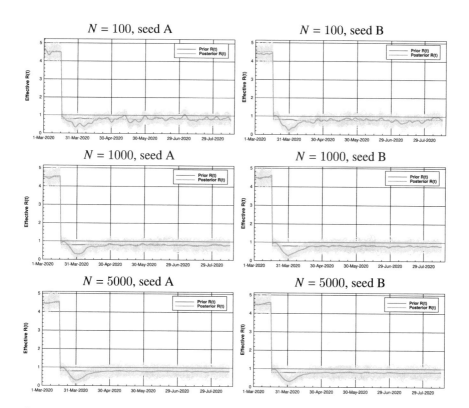

Fig. 22.4 The plots show the estimated effective reproductive number $R(t)$ for increasing ensemble sizes of $N = 100$, $N = 1000$, and $N = 5000$. The left and right columns show results from two random seeds corresponding to the plots in Fig. 22.3

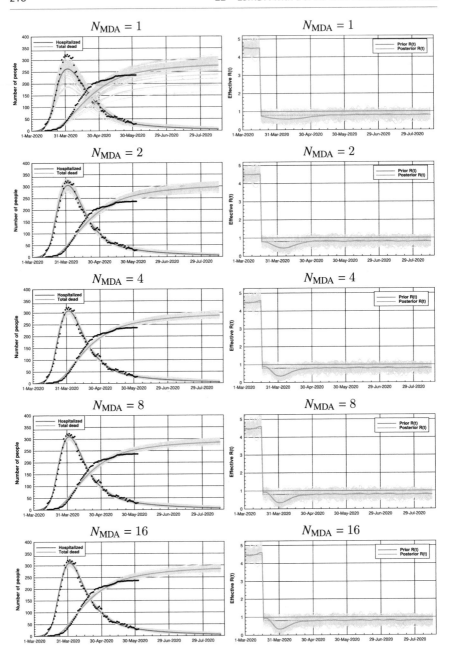

Fig. 22.5 The left plots show the estimated solution of hospitalized in red and accumulated deaths in green. The right plots show the corresponding effective reproductive number $R(t)$. From top to bottom, the results are from ESMDA with $N_{\mathrm{MDA}} = 1, 2, 4, 8,$ and 16

we increase the ensemble size to 10,000 realizations, we do not see any difference in the parameter estimates or forecasts. For this reason, Evensen et al. (2020) used 5000 realizations for all the simulations in their paper.

22.4 Sensitivity to MDA Steps

We include a sensitivity experiment to examine the convergence properties of the ESMDA algorithm with to the number of MDA steps. We expect that the accuracy of the solution will improve with the number of steps until a certain level where there is nothing more to gain. The required number of steps is, of course, dependent on the model's nonlinearity. Evensen (2018) examined the convergence of ESMDA for a simple nonlinear scalar case and obtained minimal improvement after 16–32 steps. See also the ESMDA example in Chap. 18. Figure 22.5 presents the posterior solution for deaths and hospitalized, and corresponding estimates of $R(t)$ when using ESMDA with 1, 2, 4, 8, and 16 steps. From visual inspection, it is hard to justify more than 16 steps. In Evensen et al. (2020), the authors decided to use 32 MDA steps and 5000 realizations in all simulations to ensure convergence and eliminate the possibility of sampling errors. Each experiment required a few minutes of computation on a powerful laptop with this simple model.

22.5 Summary

From the ESMDA implementation with the SEIR model, we have seen that we can accurately estimate the time-dependent $R(t)$ that ensures an excellent fit to the data up till about two weeks before the final data point. From there on, we need to specify $R(t)$ based on what we know about ongoing and planned lockdowns, the versions of mutated viruses, and the population's general behavior. There is an apparent predictive skill in the system since every action we take or intervention we implement will only influence the data a couple of weeks into the future. Thus, we only see the impact of a lockdown two weeks later. The sensitivity experiments shed light on the effects of the ensemble size on sampling errors in ensemble methods and provide insight into the convergence properties of ESMDA. This example also illustrates how we can estimate both model parameters and the forcing (controls) to constrain the model to follow the observations. Thus, we have great flexibility in formulating the problem and ensuring physically reliable solutions using data assimilation.

Final Summary

23

This final chapter provides a general summary and discussion of the data-assimilation problem that will help the reader to choose a suitable assimilation method given the problem at hand, the application's purpose, and the available time and resources. As the reader will notice, the classification below follows to a large extent the graphical representation of all the methods in Fig. 11.1 from bottom to top.

23.1   Classification of the Nonlinearity

Whether the dynamical model is linear or nonlinear, and eventually the level of non-linearity, is the primary factor to consider when selecting a suitable data-assimilation method. Thus, we will start by discussing how different levels of nonlinearity impacts the choice of the data-assimilation method. In each case, the minimal requirement is the mean or mode of the posterior pdf. However, an uncertainty estimate is crucial for scientific significance, and uncertainty estimates are essential for any real-life application.

23.1.1   Linear to Weakly-Nonlinear Systems with Gaussian Priors

We first consider the case where the system is linear Gaussian, or the forward model is weakly nonlinear. We only have to solve for the posterior mean and covariance in this case. The Kalman filter and possibly the extended Kalman filter from Chap. 6, finds the optimal solution for these problems. However, they have a drawback: the state and its *covariance matrix* must be propagated from one update step to the next, which is infeasible or impossible for high-dimensional systems. Furthermore, even with weak nonlinearity present, the linearization of the error-covariance equation

© The Author(s) 2022
G. Evensen et al., *Data Assimilation Fundamentals*,
Springer Textbooks in Earth Sciences, Geography and Environment,
https://doi.org/10.1007/978-3-030-96709-3_23

might lead to instabilities and unphysical solutions. Evensen (1994) developed the EnKF to handle the error-covariance propagation's nonlinearity and resolve the dimensionality issue. Thus, even for linear models, the EnKF can be a computationally attractive alternative, e.g., as in the linear advection example in Chap. 14.

One would use an EnKF with a large ensemble of $\mathcal{O}(100)$ realizations for high-dimensional systems. Of course, using a limited ensemble size introduces sampling errors, and we should select the ensemble size to reduce the sampling errors to an acceptable level. If a large ensemble size is not affordable, we can use an EnKF with a smaller ensemble size with the localization and inflation methods from Chap. 10. In the EnKF, the forward model equations can be nonlinear and do not need to be linearized, so the error propagation is exact.

The Kalman smoother is optimal for a smoother problem with observations distributed over an assimilation window. We have not discussed this method as we rarely use it in geosciences. A drawback is the forward and backward propagation of the error-covariance matrix over the assimilation window, making the method of little use for high-dimensional problems. For these problems, we can apply either an EnKS or an ES with a large ensemble size, or in combination with the localization and inflation methods from Chap. 10 when the ensemble size is small. If we have an adjoint model available, we can also use the representer method from Chap. 5 to solve the weak-constraint problem effectively for the mode. We can even use an ensemble of representer solutions in an RML setting (see Chap. 7) to represent the posterior uncertainty.

23.1.2 Weakly Nonlinear Systems with Gaussian Priors

We now consider model systems with weakly to modest nonlinearity and Gaussian priors. Modest nonlinearity means that the predicted measurements are monotonic functions of the state vector. This constraint eliminates cases with multiple modes in the pdf. We can distinguish situations where the adjoint of the forward model is available and situations where it is not. A general rule is that if we have access to the adjoint of the forward model, it is such a powerful tool that we should use it. In this case, the method of choice is RML sampling with adjoints, which includes EnRML, ensembles of variational methods, i.e., En3DVar, En4DVar, and the representer method. Note that the requirement of a posterior uncertainty estimate rules out a single 3DVar, or, for a smoother, a single SC-4DVar or WC-4DVar. For this reason, all operational weather prediction centers run either En4DVar, EnKF, or a single 3DVar or 4DVar augmented with an EnKF in some combination.

It is of interest to consider the complexity of implementing the different methods. The development and coding of an adjoint model can be an overwhelming task. If the adjoint of the forward model is unavailable, the recommended choices are EnRML and ESMDA, either with a large ensemble or exploring localization and inflation. Note that neither of these methods, with or without an adjoint, will provide the exact solution, but if the nonlinearity is modest, the error made is often negligible compared to other approximations in the system. For sequential data assimilation aimed for prediction, the EnRML and ESMDA may be unnecessarily computationally demanding. EnKF updates at the end of the assimilation window are straightforward and highly efficient to compute. Moreover, using existing well-tested codes and libraries for the EnKF analysis scheme, an EnKF application can be up and running in a few days. For a sequential prediction problem, it is not clear that an ensemble of 4DVar systems will perform better than a standard EnKF implementation configured to similar computational cost.

Another consideration in the choice of a method for a weakly nonlinear system, is the timing of the estimate update. We have seen that some methods update the estimate at the beginning of the window, e.g., SC-4DVar, while others estimate the update at the end of the window, e.g., EnKF. SC-4DVar assumes zero model errors, and we obtain the final solution by integrating the model over the assimilation window. The weak constraint 4DVar and the ensemble smoothers simultaneously update the model state in the whole window.

Finally, we mention that SC-4DVar and WC-4Dvar, including the representer method, are only efficient with appropriate preconditioning and reasonable estimates of the background covariance matrix. Unfortunately, no efficient preconditioning for WC-4DVar is available for the observation-space variant, the representer method, or the state- or forcing formulation. The lack of an efficient preconditioner for these weak-constraint methods is related to the vast problem size, which makes parallelization essential. However, efficient preconditioners developed for the strong-constraint problem interfere with the parallelization. Randomized preconditioners might be able to break this deadlock (see, e.g., Bousserez et al., 2020; Daužickaitė et al., 2020, 2021a, 2021b).

23.1.3 Strongly Nonlinear Systems

When the system is strongly nonlinear with multiple modes in the pdf, we must use fully nonlinear data-assimilation methods. We recommend using either particle filters or particle-flow filters. Note that we can reduce the effective nonlinearity in well-observed nonlinear systems. We saw an example in Chap. 15 where sufficiently frequent measurements managed to keep the ensemble tracking the observed attractor, avoiding bimodality. Particle filters require equivalent-weights schemes or localization, and there are applications of local particle filters with high-dimensional atmospheric problems. Particle-flow filters do not need localization by construction, and the research community has not yet fully explored this approach in detail. The only variant tested in high-dimensional systems uses kernels, and one has to find good kernels for each specific problem. However, it seems that solutions might be less sensitive to the exact kernel choice than previously thought.

23.2 Purpose of the Data Assimilation

In choosing a data-assimilation method, it is essential to consider the linearity or the Gaussianity of the system and its priors. Still, it also depends on the purpose of the data assimilation. The classification below lists several possible goals for using an assimilation system.

23.2.1 Hindcasts and Re-analyses

To analyze a system's behavior, we can assimilate all available data over a certain period into a numerical model to obtain a consistent evolution of the state. In this case, it is more important to choose a data-assimilation scheme that is computationally efficient and can incorporate a long time series of (heterogeneous) measurements than a scheme that provides accurate posterior distributions of the variables. We would still split the hindcast period into several assimilation windows and use, e.g., an EnKS or a 4DVar.

23.2.2 Prediction Systems

Let's imagine that we want to design a sequential data assimilation system used for weather predictions. What is the preferred approach? The most common and original purpose of assimilating data is to obtain the best model state at the end of an assimilation window and the best model parameters to forecast a natural system's behavior accurately. We need the solution at the end of the assimilation window to compute a new forecast. The EnKF and WC-4DVar readily provide the necessary initial conditions if the system is weakly nonlinear and has Gaussian priors. The EnKF and an ensemble of WC-4DVar systems have the added advantage of providing consistent error statistics at the initial time of the next assimilation window. These methods thereby support recursive data assimilation and allow us to initialize a prediction with quantified uncertainty.

Interestingly, the most used data-assimilation method for weather prediction is SC-4DVar, which optimizes the solution at the beginning of an assimilation window. After that, one integrates the model solution to the end of the assimilation window with the deterministic nonlinear model, where the actual weather prediction starts. The reasons for this procedure's success are twofold. First, the model state at the beginning of the assimilation window will be accurate because we have used past and future observations to update it. Second, a data-assimilation update will always push the model slightly out of its preferred balanced state, resulting in an adjustment of the model state during the first part of the forward integration, leading to less accurate forecasts. An SC-4DVar ensures that this adjustment is happening before the actual prediction.

There is a, perhaps slightly overlooked, problem when using SC-4DVar for predictions. While the SC-4DVar finds the posterior pdf's mode at the start of the assimilation window, there is no guarantee that the solution will be the mode after propagation to the end of the assimilation window where we initialize the actual forecasts (see, e.g., Van Leeuwen et al., 2015).

The pdf evolution over the assimilation window also affects smoother methods that estimate the joint pdf's mode over the whole assimilation window. For nonlinear systems, the joint pdf's mode will not be the same as the marginal pdf's mode at the end of the assimilation window (Van Leeuwen et al., 2015). And the marginal pdf's mode at the end of the assimilation window is, at least in theory, the best starting point for a prediction. This point explains one reason why data-assimilation methods that update the state at the end of the assimilation window might be preferable.

23.2.3   Uncertainty Quantification and Risk Assessment

In many applications, assessing the uncertainties in the state or parameter estimates is critical. A particular example relates to subsurface uncertainty quantification or geotechnical applications for risk assessment (e.g., Mohsan et al., 2021). EnRML, ESMDA, and EnKF are commonly used methods for these applications.

But would, in these applications, estimating the mode be enough, or do we need to sample the posterior distribution? While the mentioned methods may be easy to implement and computationally efficient, their estimates of posterior distributions in the case of the system's nonlinear behavior may be flawed. Therefore, it is essential to evaluate the nonlinearity and Gaussianity of the system, as for highly nonlinear problems, the assumptions on linearity or Gaussianity may have implications for quantifying the uncertainties or risks. The choice of method depends on whether the problem is a parameter-estimation problem that we can solve over a single assimilation window or a data assimilation problem where data become available sequentially and where we must evolve the error statistics.

23.2.4   Model Improvement and Parameter Estimation

One can distinguish parameter estimation from estimating missing physics and parameterizations. Specifically, we use data assimilation to identify missing physics, inaccurate forcing, or model errors. In specific cases, data assimilation aims not to obtain a description, forecast, or better understand a system, but the main aim is a model improvement. In some cases, data assimilation has involved estimating missing terms in model equations (Lang et al., 2016), estimating model factors (Vossepoel & Behringer, 2000), and adjusting the geometry of the model domain (Glegola et al., 2012). It is essential to apply a method that explicitly considers the model errors in such cases. We can include model errors by adding $\mathbf{q}$ to the state vectors of any ensemble or variational methods. The estimated model errors will contain both random and structural parts. The structural components point to estimated missing physics at each time step, hence at the level of the model equations, and allow for direct model improvement, contrary to comparing model forecast to observations where finding the source of the model issues is almost impossible.

23.2.5   Scenario Forecasts and Optimal Controls

In some applications, we use data assimilation to forecast a given system under a given control. It is then possible to evaluate different scenarios or optimize the control strategy. Other examples of the application of data assimilation for forecasts, scenarios, or optimal controls include the control of a producing hydrocarbon field as discussed by Jansen et al. (2009) and in Chap. 21, scenario planning for the evolution of a pandemic as presented in Chap. 22, regularizing economic processes or forecasting and controlling traffic (van Hinsbergen et al., 2012; Wang & Papageorgiou, 2005; Xie et al., 2018). In many cases, the EnKF is a practical and efficient method. In the case of parameter and control variables estimation, techniques such as EnRML or ESMDA are popular.

23.3   How to Reduce Computational Costs

Section 11.2 assumes that we can meet the necessary computational requirements, while this can be a severe issue. Present-day computer architectures are parallel, and we must explore this structure fully. Ensemble methods are naturally parallel in the forward propagation of the model state. The update in ensemble space is harder to parallelize because communication between ensemble realizations is essential. Iterative methods need the solution of one iteration before the next iteration can be processed, making these methods sequential by construction. However, research is ongoing on parallelization in the time domain. In this case, one splits the assimilation window into smaller time segments and runs parallel iterations for each, with communication only needed after each iteration.

As mentioned in Sect. 23.1, an adjoint of the forward model or the observation operator will increase accuracy and efficiency. Unfortunately, generating the adjoint of a complex forward model is a highly challenging process, often taking many years to complete. Automatic adjoint-compilers are available that need as input a forward model and a complete specification of the meaning of each symbol used in the forward model code and generate as output the tangent-linear and adjoint model codes. While these compilers are pretty sophisticated and helpful for many forward models, their output codes are (as yet) not as efficient as a human can make them. A way to improve the efficiency is to generate the adjoint code simultaneously with the forward code. Indeed, the adjoint compilers have generated some remarkably efficient codes this way, but it requires coding the forward model from scratch.

A frequent statement is that ensemble methods are too expensive, and variational methods require less computational costs and are preferable. This statement is, however, a misrepresentation of reality. First of all, variational methods need an ensemble component for uncertainty quantification. But the most crucial argument is that each iteration of a 4DVar contains one forward tangent-linear and one adjoint integration over the assimilation window. This computation corresponds roughly to integrating two ensemble realizations. Hence, 50 iterations correspond to an ensemble size of approximately 100.

Furthermore, the linearizations in the iterative methods lead to additional terms in the model equations. A tangent linear or adjoint integration is two to four times more expensive than a nonlinear model integration used in an ensemble method. However, there are ways of compensating for this additional cost in variational methods. We can reduce the resolution of the models used in the iterations or simplify the nonlinear forward model's nonlinear parameterizations. Both approaches introduce additional approximations. It seems that variational and ensemble methods have similar costs for similar accuracy in practice.

23.4   What Will the Future Hold?

Of course, it is almost impossible to predict the future, even given all the data we have, because human evolution is perhaps more chaotic than nature. Nonetheless, we can pinpoint some trends that might have some momentum, hence predictive power.

First, there is an increasing trend of pushing for nonlinear data-assimilation methods. For instance, the ever-increasing resolution in weather prediction models resolves turbulent features in the atmospheric boundary layer. We do not have enough observations to avoid the development of strongly non-Gaussian pdfs. Indeed, a growing number of scientists within applied and even pure mathematics are among those pushing the boundaries.

Another trend is the incorporation of machine learning in data assimilation. Examples include using machine learning to make models more efficient, accelerating data-assimilation methods, and replacing parts of the data-assimilation process. It is hard to predict where this is heading. An important lesson seems to be that, e.g., neural networks from image processing are not directly applicable for data assimilation, and we need particular machine architectures to make real progress. Physically realistic predictions seem achievable by building strong model constraints into the machine-learning cost function, bringing machine learning closer to variational methods where model constraints incorporate prior knowledge.

A general weakness of machine learning is its inability to estimate uncertainty. That will need to change if machine learning will be an alternative for data assimilation for real-world applications in geosciences. Indeed, many ideas from data-assimilation on uncertainty quantification are starting to find their way in the machine learning literature, either being actively brought in or reinvented. Unfortunately, not all application areas are fully aware of other methods, and some practitioners reinvent techniques that have been mainstream in different research fields for decades.

Whatever the future, it will most likely result from the collaborations of many scientists from many different disciplines. If we keep the data-assimilation field as collaborative as it is now, the future is bright.

References

Aanonsen, S. I., Naevdal, G., Oliver, D. S., Reynolds, A. C., & Valles, B. (2009). Ensemble Kalman filter in reservoir engineering—A review. *SPE Journal, 14*(3), 393–412. https://doi.org/10.21188/117274-PA.

Ades, M., & Van Leeuwen, P. J. (2013). An exploration of the equivalent weights particle filter. *Quarterly Journal of the Royal Meteorological Society, 139*, 820–840. https://doi.org/10.1002/qj.1995.

Ades, M., & Van Leeuwen, P. J. (2015a). The equivalent-weights particle filter in a high-dimensional system. *Quarterly Journal of the Royal Meteorological Society, 141*, 484–503. https://doi.org/10.1002/qj.2370.

Ades, M., & Van Leeuwen, P. J. (2015b). The effect of the equivalent-weights particle filter on model balances in a primitive equation model. *Monthly Weather Review, 143*. https://doi.org/10.1175/MWR-D-14-00050.1.

Amezcua, J., Goodliff, M., & Van Leeuwen, P. J. (2017). A weak-constraint 4DEnsembleVar. Part I: Formulation and simple model experiments. *Tellus.* https://doi.org/10.1080/16000870.2016.1271564.

Anderson, J. L. (2003). A local least squares framework for ensemble filtering. *Monthly Weather Review, 131*, 634–642. https://doi.org/10.1175/1520-0493(2003)131<0634:ALLSFF>2.0.CO;2.

Anderson, J. L. (2007a). An adaptive covariance inflation error correction algorithm for ensemble filters. *Tellus Series A, 59*, 210–224. https://doi.org/10.1111/j.1600-0870.2006.00216.x.

Anderson, J. L. (2007b). Exploring the need for localization in the ensemble data assimilation using a hierarchical ensemble filter. *Physica D, 230*, 99–111. https://doi.org/10.1016/j.physd.2006.02.011.

Anderson, J. L. (2009). Spatially and temporally varying adaptive covariance inflation for ensemble filters. *Tellus Series A, 61*, 72–83. https://doi.org/10.1111/j.1600-0870.2008.00361.x.

Anderson, J. L., & Anderson, S. L. (1999). A Monte Carlo implementation of the nonlinear filtering problem to produce ensemble assimilations and forecasts. *Monthly Weather Review, 127*, 2741–2758. https://doi.org/10.1175/1520-0493(1999)127<2741:AMCIOT>2.0.CO;2.

© The Editor(s) (if applicable) and The Author(s) 2022
G. Evensen et al., *Data Assimilation Fundamentals*,
Springer Textbooks in Earth Sciences, Geography and Environment,
https://doi.org/10.1007/978-3-030-96709-3

Asch, M., Bocquet, M., & Nodet, M. (2017). Data assimilation: Methods, algorithms, and applications. *SIAM, 10*(1137/1), 9781611974546.

Bain, A., & Crisan, D. (2009). *Fundamentals of Stochastic Filtering*. Springer. ISBN 978-0-387-76895-3.

Banerjee, A., van Dinther, Y., & Vossepoel, F. C. (2022). On parameter bias in earthquake sequence models using data assimilation. *Submitted to Nonlinear Processes in Geophysics*.

Bannister, R. N. (2008). Modelling the forecast error covariance statistics: A review of forecast error covariance statistics in atmospheric variational data assimilation II. *Quarterly Journal of the Royal Meteorological Society, 134,* 1971–1996. https://doi.org/10.1002/qj.340.

Bannister, R. N. (2017). Review article: A review of operational methods of variational and ensemble-variational data assimilation. *Quarterly Journal of the Royal Meteorological Society, 143,* 607–633. https://doi.org/10.1002/qj.2982.

Bengtsson, T., Snyder, C., & Nychka, D. (2003). Toward a nonlinear ensemble filter for high-dimensional systems. *Journal of Geophysical Research, 108,* 8775–8785. https://doi.org/10.1029/2002JD002900.

Bennett, A. F. (1992). *Inverse Methods in Physical Oceanography*. Cambridge University Press. https://doi.org/10.1017/CBO9780511600807.

Bennett, A. F. (2002). *Inverse Modeling of the Ocean and Atmosphere*. Cambridge University Press. https://doi.org/10.1017/CBO9780511535895.

Bennett, A. F., & McIntosh, P. C. (1982). Open ocean modeling as an inverse problem: Tidal theory. *Journal of Physical Oceanography, 12,* 1004–1018. https://doi.org/10.1175/1520-0485(1982)012<1004:OOMAAI>2.0.CO;2.

Bennett, A. F., Chua, B. S., & Leslie, L. M. (1997). Generalized inversion of a global numerical weather prediction model, II: Analysis and implementation. *Meteorology and Atmospheric Physics, 62*(3), 129–140. https://doi.org/10.1007/BF01029698.

Bishop, C. H., & Hodyss, D. (2007). Flow-adaptive moderation of spurious ensemble correlations and its use in ensemble-based data assimilation. *Quarterly Journal of the Royal Meteorological Society, 133,* 2029–2044. https://doi.org/10.1002/qj.169.

Bishop, C. H., Etherton, B. J., & Majumdar, S. J. (2001). Adaptive sampling with the ensemble transform Kalman filter. Part I: Theoretical aspects. *Monthly Weather Review, 129,* 420–436. https://doi.org/10.1175/1520-0493(2001)129<0420:ASWTET>2.0.CO;2.

Blackwood, J. C., & Childs, L. M. (2018). An introduction to compartmental modeling for the budding infectious disease modeler. *Letters in Biomathematics, 5,* 195–221. https://doi.org/10.1080/23737867.2018.1509026.

Bocquet, M. (2015). Localization and the iterative ensemble Kalman smoother. *Quarterly Journal of the Royal Meteorological Society, 141.* https://doi.org/10.1002/qj.2711.

Bocquet, M., Raanes, P. N., & Hannart, A. (2015). Expanding the validity of the ensemble Kalman filter without the intrinsic need for inflation. *Nonlinear Processes in Geophysics, 22*(6), 645–662. https://doi.org/10.5194/npg-22-645-2015.

Bousserez, N., Guerrette, J. J., & Henze, D. K. (2000). Enhanced parallelization of the incremental 4D-Var data assimilation algorithm using the randomized incremental optimal technique. *Quarterly Journal of the Royal Meteorological Society, 146.* https://doi.org/10.1002/qj.3740.

Bowers, C. M., Price, J. F., Weller, R. A., & Briscoe, M. G. (1986). Data tabulations and analysis of diurnal sea surface temperature variability observed at LOTUS. Woods Hole Oceanographic Institution (vol. 5). https://doi.org/10.1575/1912/7867.

Brusdal, K., Brankart, J. M., Halberstadt, G., Evensen, G., Brasseur, P., Van Leeuwen, P. J., Dombrowsky, E., & Verron, J. (2003). A demonstration of ensemble-based assimilation methods with a layered OGCM from the perspective of operational ocean forecasting systems. *Journal of Marine Systems, 40–41,* 253–289. https://doi.org/10.1016/S0924-7963(03)00021-6.

Buehner, M. (2005). Ensemble-derived stationary and flow-dependent background-error covariances: Evaluation in a quasi-operational NWP setting. *Quarterly Journal of the Royal Meteorological Society, 131,* 1013–1043. https://doi.org/10.1256/qj.04.15.

Bui-Thanh, T. (2021). The optimality of Bayes' theorem. *SIAM News, 54*(6), 1–4. https://sinews.siam.org/Details-Page/the-optimality-of-bayes-theorem.

Burgers, G., Van Leeuwen, P. J., & Evensen, G. (1998). Analysis scheme in the ensemble Kalman filter. *Monthly Weather Review, 126,* 1719–1724. https://doi.org/10.1175/1520-0493(1998)126<1719:ASITEK>2.0.CO;2.

Burridge, R., & Knopoff, L. (1967). Model and theoretical seismicity. *Bulletin of the Seismological Society of America, 57*(3), 341–371. https://doi.org/10.1785/BSSA0570030341.

Cao, H., & Zhou, Y. (2012). The discrete age-structured SEIT model with application to tuberculosis transmission in China. *Mathematical and Computer Modelling, 55,* 385–395. https://doi.org/10.1016/j.mcm.2011.08.017.

Carrassi, A., Bocquet, M., Bertino, L., & Evensen, G. (2018). Data assimilation in the geosciences: An overview on methods, issues and perspectives. *Wires Climate Change, 9*(5), 50. https://doi.org/10.1002/wcc.535.

Chen, Y., & Oliver, D. S. (2012). Ensemble randomized maximum likelihood method as an iterative ensemble smoother. *Mathematical Geosciences, 44,* 1–26. https://doi.org/10.1007/s11004-011-9376-z.

Chen, Y., & Oliver, D. S. (2013). Levenberg-Marquardt forms of the iterative ensemble smoother for efficient history matching and uncertainty quantification. *Computational Geosciences, 17,* 689–703. https://doi.org/10.1007/s10596-013-9351-5.

Chen, Y., & Oliver, D. S. (2017). Localization and regularization for iterative ensemble smoothers. *Computational Geosciences, 21,* 13–30. https://doi.org/10.1007/s10596-016-9599-7.

Chorin, A. J., & Tu, X. (2009). Implicit sampling for particle filters. *PNAS, 106,* 17249–17254. https://doi.org/10.1073/pnas.0909196106.

Chorin, A. J., Morzfeld, M., & Tu, X. (2010). Implicit particle filters for data assimilation. *Communications in Applied Mathematics and Computational Science.*

Courtier, P. (1997). Dual formulation of variational assimilation. *Quarterly Journal of the Royal Meteorological Society, 123,* 2449–2461. https://doi.org/10.1002/qj.49712354414.

Daley, R. (1991). *Atmospheric Data Analysis.* Cambridge University Press. ISBN 9780521458252.

Daum, F., & Huang, J. (2011). Particle flow for nonlinear filters. In *Proceedings of the IEEE International Conference on Acoustics, Speech, and Signal Processing,* ICASSP 2011, May 22–27, 2011, Prague Congress Center, Prague, Czech Republic (pp. 5920–5923). https://doi.org/10.1109/ICASSP.2011.5947709.

Daum, F., & Huang, J. (2013). Particle flow for nonlinear filters, Bayesian decisions and transport. In *Proceedings of the 16th International Conference on Information Fusion, FUSION 2013,* Istanbul, Turkey, July 9–12, 2013 (pp. 1072–1079). http://ieeexplore.ieee.org/document/6641115/.

Daum, F., Huang, J., & Noushin, A. (2018). New theory and numerical results for Gromov's method for stochastic particle flow filters. In *2018 21st International Conference on Information Fusion (FUSION)* (pp. 108–115). https://doi.org/10.23919/ICIF.2018.8455287.

Daužickaitė, I., Lawless, A. S., Scott, J., & Van Leeuwen, P. J. (2020). Spectral estimates for saddle point matrices arising in weak constraint four-dimensional variational data assimilation. *Numerical Linear Algebra with Applications.* https://doi.org/10.1002/nla.2313.

Daužickaitė, I., Lawless, A. S., Scott, J., & Van Leeuwen, P. J. (2021a). Randomised preconditioning for the forcing formulation of weak constraint 4D-Var. *Quarterly Journal of the Royal Meteorological Society.* https://doi.org/10.1002/qj.4151.

Daužickaitė, I., Lawless, A. S., Scott, J., & Van Leeuwen, P. J. (2021b). On time-parallel preconditioning for the state formulation of incremental weak constraint 4D-Var. *Quarterly Journal of the Royal Meteorological Society.* https://doi.org/10.1002/qj.4140.

Derber, J. C. (1989). A variational continuous assimilation technique. *Monthly Weather Review, 174*, 2437–2446. https://doi.org/10.1175/1520-0493(1989)117<2437:AVCAT>2.0.CO;2.

Egbert, G. D., Bennett, A. F., & Foreman, M. G. G. (1994). TOPEX/POSEIDON tides estimated using a global inverse model. *Journal of Geophysical Research, 99*(C12), 24821–24852. https://doi.org/10.1029/94JC01894.

Eknes, M., & Evensen, G. (1997). Parameter estimation solving a weak constraint variational formulation for an Ekman model. *Journal of Geophysical Research, 102*(C6), 12479–12491. https://doi.org/10.1029/96JC03454.

El Moselhy, T. A., & Marzouk, Y. M. (2012). Bayesian inference with optimal maps. *Journal of Computational Physics, 231*, 7815–7850. https://doi.org/10.1016/j.jcp.2012.07.022.

Emerick, A. A. (2018). Deterministic ensemble smoother with multiple data assimilation as an alternative for history matching seismic data. *Computational Geosciences, 22*, 10. https://doi.org/10.1007/s10596-018-9745-5.

Emerick, A. A., & Reynolds, A. C. (2013). Ensemble smoother with multiple data assimilation. *Computers and Geosciences, 55*, 3–15. https://doi.org/10.1016/j.cageo.2012.03.011.

Erickson, B., Birnir, B., & Lavallée, D. (2011). Periodicity, chaos and localization in a Burridge-Knopoff model of an earthquake with rate-and-state friction. *Geophysical Journal International, 187*, 178–198. https://doi.org/10.1111/j.1365-246X.2011.05123.x.

Evensen, G. (1992). Using the extended Kalman filter with a multilayer quasi-geostrophic ocean model. *Journal of Geophysical Research, 97*(C11), 17905–17924. https://doi.org/10.1029/92JC01972.

Evensen, G. (1994). Sequential data assimilation with a nonlinear quasi-geostrophic model using Monte Carlo methods to forecast error statistics. *Journal of Geophysical Research, 99*(C5), 10143–10162. https://doi.org/10.1029/94JC00572.

Evensen, G. (1997). Advanced data assimilation for strongly nonlinear dynamics. *Monthly Weather Review, 125*, 1342–1354. https://doi.org/10.1175/1520-0493(1997)125<1342:ADAFSN>2.0.CO;2.

Evensen, G. (2003). The ensemble Kalman filter: Theoretical formulation and practical implementation. *Ocean Dynamics, 53*, 343–367. https://doi.org/10.1007/s10236-003-0036-9.

Evensen, G. (2004). Sampling strategies and square root analysis schemes for the EnKF. *Ocean Dynamics, 54*, 539–560. https://doi.org/10.1007/s10236-004-0099-2.

Evensen, G. (2009a). The ensemble Kalman filter for combined state and parameter estimation. *IEEE Control Systems Magazine, 29*(3), 83–104. https://doi.org/10.1109/MCS.2009.932223.

Evensen, G. (2009b). *Data Assimilation: The Ensemble Kalman Filter* (2nd ed.) Springer. https://doi.org/10.1007/978-3-642-03711-5.

Evensen, G. (2018). Analysis of iterative ensemble smoothers for solving inverse problems. *Computational Geosciences, 22*(3), 885–908. https://doi.org/10.1007/s10596-018-9731-y.

Evensen, G. (2019). Accounting for model errors in iterative ensemble smoothers. *Computational Geosciences, 23*(4), 761–775. https://doi.org/10.1007/s10596-019-9819-z.

Evensen, G. (2021). Formulating the history matching problem with consistent error statistics. *Computational Geosciences, 25*, 945–970. https://doi.org/10.1007/s10596-021-10032-7.

Evensen, G., & Eikrem, K. S. (2018). Strategies for conditioning reservoir models on rate data using ensemble smoothers. *Computational Geosciences, 22*(5), 1251–1270. https://doi.org/10.1007/s10596-018-9750-8.

Evensen, G., & Van Leeuwen, P. J. (2000). An ensemble Kalman smoother for nonlinear dynamics. *Monthly Weather Review, 128*, 1852–1867. https://doi.org/10.1175/1520-0493(2000)128<1852:AEKSFN>2.0.CO;2.

Evensen, G., Hove, J., Meisingset, H. C., Reiso, E., Seim, K. S., & Espelid, Ø. (2007). Using the EnKF for assisted history matching of a North Sea reservoir model. In *SPE Conference Paper*. https://doi.org/10.2118/106184-MS.

Evensen, G., Raanes, P. N., Stordal, A. S., & Hove, J. (2019). Efficient implementation of an iterative ensemble smoother for data assimilation and reservoir history matching. *Frontiers in Applied Mathematics and Statistics, 5,* 47. https://doi.org/10.3389/fams.2019.00047.

Evensen, G., Amezcua, J., Bocquet, M., Carrassi, A., Farchi, A., Fowler, A., Houtekamer, P. L., Jones, C. K., de Moraes, R. J., Pulido, M., Sampson, C., & Vossepoel, F. C. (2020). An international initiative of predicting the sars-cov-2 pandemic using ensemble data assimilation. *Foundations of Data Science, 65.* https://doi.org/10.3934/fods.2021001.

Fertig, E. J., Hunt, B. R., Ott, E., & Szunyogh, I. (2007). Assimilating non-local observations with a local ensemble Kalman filter. *Tellus Series A, 59,* 719–730. https://doi.org/10.1111/j.1600-0870.2007.00260.x.

Fichtner, A. (2011). *Full Seismic Waveform Modelling and Inversion.* Springer. ISBN 978-3-642-15807-0.

Fisher, M., Trémolet, Y., Auvinen, H., Tan, D., & Poli, P. (2011). Weak-constraint and long-window 4D-Var. *ECMWF Technical Memorandum, 655,* 1–47. https://doi.org/10.21957/9ii4d4dsq.

Fletcher, S. J. (2017). *Data Assimilation for the Geosciences.* Elsevier. ISBN 978-0-12-804444-5.

Furrer, R., & Bengtsson, T. (2007). Estimation of high-dimensional prior and posterior covariance matrices in Kalman filter variants. *Journal of Multivariate Analysis, 98*(2), 227–255. https://doi.org/10.1016/j.jmva.2006.08.003.

Gaspari, G., & Cohn, S. (1999). Construction of correlation functions in two and three dimensions. *Quarterly Journal of the Royal Meteorological Society, 125,* 723–757. https://doi.org/10.1002/qj.49712555417.

Glegola, M., Ditmar, P., Hanea, R. G., Eiken, O., Vossepoel, F. C., Arts, R. J., & Klees, R. (2012). History matching time-lapse surface-gravity and well-pressure data with ensemble smoother for estimating gas field aquifer support-A 3D numerical study. *SPE Journal, 17*(4), 966–980. https://doi.org/10.2118/161483-PA.

Greybush, S. J., Kalnay, E., Miyoshi, T., Ide, K., & Hunt, B. R. (2011). Balance and ensemble Kalman filter localization techniques. *Monthly Weather Review, 139,* 511–522. https://doi.org/10.1175/2010MWR3328.1.

Gu, Y., & Oliver, D. S. (2005). History matching of the PUNQ-S3 reservoir model using the ensemble Kalman filter. *SPE Journal, 10*(2), 217–224. https://doi.org/10.2118/89942-PA.

Hamill, T. M., Whitaker, J. S., & Snyder, C. (2001). Distance-dependent filtering of background error covariance estimates in an ensemble Kalman filter. *Monthly Weather Review, 129,* 2776–2790. https://doi.org/10.1175/1520-0493(2001)129<2776:DDFOBE>2.0.CO;2.

Haugen, V. E., & Evensen, G. (2002). Assimilation of SLA and SST data into an OGCM for the Indian ocean. *Ocean Dynamics, 52,* 133–151. https://doi.org/10.1007/s10236-002-0014-7.

Haugen, V. E., Nævdal, G., Natvik, L.-J., Evensen, G., Berg, A., & Flornes, K. (2008). History matching using the ensemble Kalman filter on a north sea field case. *SPE Journal, 13,* 382–391. https://doi.org/10.2118/102430-PA.

Hodyss, D., & Nichols, N. K. (2015). The error of representation: Basic understanding. *Tellus Series A, 67.* https://doi.org/10.3402/tellusa.v67.24822.

Hoffman, R. N., & Kalnay, E. (1983). Lagged average forecasting, an alternative to Monte Carlo forecasting. *Tellus A: Dynamic Meteorology and Oceanography, 35*(2), 100–118. https://doi.org/10.3402/tellusa.v35i2.11425.

Houtekamer, P. L., & Mitchell, H. L. (2001). A sequential ensemble Kalman filter for atmospheric data assimilation. *Monthly Weather Review, 129,* 123–137. https://doi.org/10.1175/1520-0493(2001)129<0123:ASEKFF>2.0.CO;2.

Houtekamer, P. L., & Zhang, F. (2016). Review of the ensemble Kalman filter for atmospheric data assimilation. *Monthly Weather Review, 144,* 4489–4533. https://doi.org/10.1175/MWR-D-15-0440.1.

Hu, C.-C., & Van Leeuwen, P. J. (2021). A particle flow filter for high-dimensional system applications. *Quarterly Journal of the Royal Meteorological Society*. https://doi.org/10.1002/qj.4028.

Hunt, B. R., Kostelich, E. J., & Szunyogh, I. (2007). Efficient data assimilation for spatiotemporal chaos: A local ensemble transform Kalman filter. *Physica D, 230*, 112–126. https://doi.org/10.1016/j.physd.2006.11.008.

Isaksen, L., Bonavita, M., Buizza, R., Fisher, M., Haseler, J., Leutbecher, M., & Raynaud, L. (2010). Ensemble of data assimilations at ECMWF. In *Technical Memorandum 636, ECMWF, Reading, UK*. https://www.ecmwf.int/node/10125.

Janjić, T., Bormann, N., Bocquet, M., Carton, J. A., Cohn, S. E., Dance, S. L., Losa, S. N., Nichols, N. K., Potthast, R., Waller, J. A., & Weston, P. (2018). On the representation error in data assimilation. *Quarterly Journal of the Royal Meteorological Society, 144*, 1257–1278. https://doi.org/10.1002/qj.3130.

Jansen, J.-D., Brouwer, R., & Douma, S. G. (2009). Closed loop reservoir management. In *SPE Conference Paper*. https://doi.org/10.2118/119098-MS.

Jazwinski, A. H. (1970). *Stochastic Processes and Filtering Theory*. Academic Press. ISBN 9780080960906.

Kalman, R. E. (1960). A new approach to linear filter and prediction problems. *Journal of Basic Engineering, 82*, 35–45. https://doi.org/10.1115/1.3662552.

Kalnay, E. (2002). *Atmospheric Modeling, Data Assimilation and Predictability*. Cambridge University Press. https://doi.org/10.1017/CBO9780511802270.

Kitagawa, G. (1996). Monte-Carlo filter and smoother for non-Gaussian non-linear state-space models. *Journal of Computational and Graphical Statistics*.

Kitanidis, P. K. (1995). Quasi-linear geostatistical theory for inversing. *Water Resources Research, 31*(10), 2411–2419. https://doi.org/10.1029/95WR01945.

Lang, M., Van Leeuwen, P. J., & Browne, P. (2016). A systematic method of parameterization estimation using data assimilation. *Tellus Series A*. https://doi.org/10.3402/tellusa.v68.29012.

Law, K., Stuart, A. M., & Zygalakis, K. (2015). *Data Assimilation: A Mathematical Introduction, volume 62 of Texts in Applied Mathematics BT—Data Assimilation: A Mathematical Introduction*. Springer. https://doi.org/10.1007/978-3-319-20325-6.

Lawless, A. S., Gratton, S., & Nichols, N. K. (2005). Approximate iterative methods for variational data assimilation. *International Journal for Numerical Methods in Fluids, 47*. https://doi.org/10.1002/fld.851.

Le Dimet, F.-X., & Talagrand, O. (1986). Variational algorithms for analysis and assimilation of meteorological observations: Theoretical aspects. *Tellus Series A, 38*, 97–110. https://doi.org/10.1111/j.1600-0870.1986.tb00459.x.

Leeuwenburgh, O. (2005). Assimilation of along-track altimeter data in the Tropical Pacific region of a global OGCM ensemble. *Quarterly Journal of the Royal Meteorological Society, 131*, 2455–2472. https://doi.org/10.1256/qj.04.146.

Lewis, J. M., & Derber, J. C. (1985). The use of adjoint equations to solve a variational adjustment problem with advective constraints. *Tellus Series A, 37*, 309–322. https://doi.org/10.3402/tellusa.v37i4.11675.

Lewis, J. M., Lakshmivarahan, S., & Dhall, S. K. (2006). *Dynamic Data Assimilation: A Least Squares Approach*. Cambridge University Press. ISBN 978-0-521-85155-8.

Liu, Q., & Wang, D. (2016). Stein variational gradient descent: A general purpose Bayesian inference algorithm. In *NIPS 2016*. arXiv:1608.04471.

Liu, Y., Weerts, A. H., Clark, M., Hendricks Franssen, H. J., Kumar, S., Moradkhani, H., Seo, D. J., Schwanenberg, D., Smith, P., Van Dijk, A. I. J. M., Van Velzen, N., He, M., Lee, H., Noh, S. J., Rakovec, O., & Restrepo, P. (2012). Advancing data assimilation in operational hydrologic forecasting: Progresses, challenges, and emerging opportunities. *Hydrology and Earth System Sciences, 16*(10), 3863–3887. https://doi.org/10.5194/hess-16-3863-2012.

Lorenc, A. C. (1988). Optimal nonlinear objective analysis. *Quarterly Journal of the Royal Meteorological Society, 114,* 205–240. https://doi.org/10.1002/qj.49711447911.

Lorenz, E. N. (1963). Deterministic nonperiodic flow. *Journal of the Atmospheric Sciences, 20,* 130–141. https://doi.org/10.1175/1520-0469(1963)020<0130:DNF>2.0.CO;2.

Lu, J., Lu, Y., & Nolen, J. (2019). Scaling limit of the Stein variational gradient descent: The mean field regime. *IAM Journal on Mathematical Analysis.* https://doi.org/10.1137/18M1187611.

Luo, X., & Bhakta, T. (2020). Automatic and adaptive localization for ensemble-based history matching. *Journal of Petroleum Science and Engineering, 184,* 106559. ISSN 0920-4105. https://doi.org/10.1016/j.petrol.2019.106559.

Luo, X., Lorentzen, R. J., Valestrand, R., & Evensen, G. (2019). Correlation-based adaptive localization for ensemble-based history matching: Applied to the Norne field case study. *SPE Reservoir Evaluation & Engineering, 22*(3), 1084–1109. https://doi.org/10.2118/191305-PA.

Majda, A. J., & Harlim, J. (2012). *Filtering Complex Turbulent Systems.* Cambridge University Press. ISBN 978-1-107-01666-8.

Marotzke, J., Giering, R., Zhang, K. Q., Stammer, D., Hill, C., & Lee, T. (1999). Construction of the adjoint MIT ocean general circulation model and application to Atlantic heat transport sensitivity. *Journal of Geophysical Research, 104*(C12), 29529–29547. https://doi.org/10.1029/1999JC900236.

McLaughlin, D. (1995). Recent developments in hydrologic data assimilation. *Reviews of Geophysics, 33*(S2), 977–984. https://doi.org/10.1029/95RG00740.

Miller, R. N., Ghil, M., & Gauthiez, F. (1994). Advanced data assimilation in strongly nonlinear dynamical systems. *Journal of the Atmospheric Sciences, 51,* 1037–1056. https://doi.org/10.1175/1520-0469(1994)051<1037:ADAISN>2.0.CO;2.

Mohsan, M., Vardon, P. J., & Vossepoel, F. C. (2021). On the use of different constitutive models in data assimilation for slope stability. *Computers and Geotechnics, 138,* 104332. https://doi.org/10.1016/j.compgeo.2021.104332.

Morzfeld, M., Tu, X., Atkins, E., & Chorin, A. J. (2012). A random map implementation of implicit filters. *Journal of Computational Physics, 231,* 2049–2066. https://doi.org/10.1016/j.jcp.2011.11.022.

Navon, I. M., & Legler, D. M. (1987). Conjugate-gradient methods for large-scale minimization in meteorology. *Monthly Weather Review, 115*(8), 1479–1502. https://doi.org/10.1175/1520-0493(1987)115<1479:CGMFLS>2.0.CO;2.

Neal, R. M. (1996). Sampling from multimodal distributions using tempered transitions. *Statistics and Computing, 6*(4), 353–366. https://doi.org/10.1007/BF00143556.

Neto, G. M. S., Soares, R. V., Evensen, G., Davolioa, A., & Schiozer, D. J. (2021). Subspace ensemble randomized maximum likelihood with local analysis for time-lapse-seismic-data assimilation. *SPE Journal, 26*(2), 1011–1031. https://doi.org/10.2118/205029-PA.

Nolet, G. (2008). *A Breviary of Seismic Tomography.* Cambridge University Press. ISBN 978-0-521-88244-6.

Oliver, D. S., & Chen, Y. (2011). Recent progress on reservoir history matching: A review. *Computational Geosciences, 15*(1), 185–221. https://doi.org/10.1007/s10596-010-9194-2.

Oliver, D. S., He, N., & Reynolds, A. C. (1996). Conditioning permeability fields to pressure data. In *ECMOR V-5th European Conference on the Mathematics of Oil Recovery* (p. 11). https://doi.org/10.3997/2214-4609.201406884.

Oliver, D. S., Reynolds, A. C., & Liu, N. (2008). *Inverse Theory for Petroleum Reservoir Characterization and History Matching.* Cambridge University Press. https://doi.org/10.1017/CBO9780511535642.

Ott, E., Hunt, B. R., Szunyogh, I., Zimin, A. V., Kostelich, E. J., Corazza, M., Kalnay, E., Patil, D. J., & Yorke, J. A. (2004). A local ensemble Kalman filter for atmospheric data assimilation. *Tellus Series A, 56A,* 415–428. https://doi.org/10.3402/tellusa.v56i5.14462.

Papadakis, N., Memin, E., Cuzol, A., & Gengembre, N. (2010). Data assimilation with the weighted ensemble Kalman filter. *Tellus A: Dynamic Meteorology and Oceanography, 62.* https://doi.org/10.1111/j.1600-0870.2010.00461.x.

Peters, E., Arts, R. J., Brouwer, G. K., Geel, C. R., Cullick, S., Lorentzen, R. J., Chen, Y., Dunlop, K. N. B., Vossepoel, F. C., Xu, R., Sarma, P., Alhutali, A. H., & Reynolds, A. C. (2010). Results of the Brugge benchmark study for flooding optimization and history matching. *SPE Reservoir Evaluation & Engineering, 13,* 391–405. https://doi.org/10.2118/119094-PA.

Pinheiro, F. R., Van Leeuwen, P. J., & Geppert, G. (2019a). Efficient nonlinear data assimilation using synchronisation in a particle filter. *Quarterly Journal of the Royal Meteorological Society.* https://doi.org/10.1002/qj.3576.

Pinheiro, F. R., Van Leeuwen, P. J., & Parlitz, U. (2019b). An ensemble framework for time-delayed synchronisation. *Quarterly Journal of the Royal Meteorological Society.* https://doi.org/10.1002/qj.3204.

Pires, C., Vautard, R., & Talagrand, O. (1996). On extending the limits of variational assimilation in nonlinear chaotic systems. *Tellus Series A, 48,* 96–121. https://doi.org/10.1034/j.1600-0870.1996.00006.x.

Poterjoy, J. (2016). A localized particle filter for high-dimensional nonlinear systems. *Monthly Weather Review, 144,* 59–76. https://doi.org/10.1175/MWR-D-15-0163.1.

Poterjoy, J., & Anderson, J. L. (2016). Efficient assimilation of simulated observations in a high-dimensional geophysical system using a localized particle filter. *Monthly Weather Review, 144.* (*Nonlinear Processes Geophysics*). https://doi.org/10.1175/MWR-D-15-0163.1.

Potthast, R., Walter, A., & Rhodin, A. (2019). A localized adaptive particle filter within an operational NWP framework. *Monthly Weather Review, 147.* https://doi.org/10.1175/MWR-D-18-0028.1.

Pulido, M., & Van Leeuwen, P. J. (2019). Sequential Monte Carlo with kernel embedded mappings: The mapping particle filter. *Journal of Computational Physics, 396,* 400–415. https://doi.org/10.1016/j.jcp.2019.06.060.

Raanes, P. N., Stordal, A. S., & Evensen, G. (2019). Revising the stochastic iterative ensemble smoother. *Nonlinear Processes in Geophysics, 26,* 325–338. https://doi.org/10.5194/npg-2019-10.

Reich, S. (2011). A dynamical systems framework for intermittent data assimilation. *BIT Numerical Mathematics, 51,* 235–249. https://doi.org/10.1007/s10543-010-0302-4.

Reich, S. (2012). A Gaussian mixture ensemble transform filter. *Quarterly Journal of the Royal Meteorological Society, 138,* 222–233. https://doi.org/10.1002/qj.898.

Reich, S. (2019). Data assimilation: The Schroedinger perspective. *Acta Numerica.* https://doi.org/10.1017/S0962492919000011.

Reich, S., & Cotter, C. J. (2015). *Probabilistic Forecasting and Bayesian Data Assimilation.* Cambridge University Press. https://doi.org/10.1017/CBO9781107706804.

Reynolds, A. C., Zafari, M., & Li, G.: Iterative forms of the Ensemble Kalman Filter. In *Proceedings of the ECMOR X—10th European Conference on the Mathematics of Oil Recovery.* European Association of Geoscientists & Engineers. https://doi.org/10.3997/2214-4609.201402496.

Ruina, A. (1983). Slip instability and state variable friction laws. *Journal of Geophysical Research: Solid Earth, 88*(B12), 10359–10370. https://doi.org/10.1029/JB088iB12p10359.

Sacher, W., & Bartello, P. (2008). Sampling errors in ensemble Kalman filtering. Part I: Theory. *MWR, 136,* 3035–3049. https://doi.org/10.1175/2007MWR2323.1.

Sakov, P., & Bertino, L. (2011). Relation between two common localization methods for the EnKF. *Computational Geosciences, 15,* 225–237. https://doi.org/10.1007/s10596-010-9202-6.

Sakov, P., Evensen, G., & Bertino, L. (2010). Asynchronous data assimilation with the EnKF. *Tellus Series A, 62A,* 24–29. https://doi.org/10.1111/j.1600-0870.2009.00417.x.

Sakov, P., Oliver, D. S., & Bertino, L. (2012). An iterative EnKF for strongly nonlinear systems. *Monthly Weather Review, 140,* 1988–2004. https://doi.org/10.1175/MWR-D-11-00176.1.

Sasaki, Y. (1970a). Some basic formalisms in numerical variational analyses. *Monthly Weather Review, 98,* 875–883. https://doi.org/10.1175/1520-0493(1970)098<0875:SBFINV>2.3.CO;2.

Sasaki, Y. (1970b). Numerical variational analyses with weak constraint and application to surface analyses of severe storm gust. *Monthly Weather Review, 98*(12), 899–910. https://doi.org/10. 1175/1520-0493(1970)098<0899:NVAWWC>2.3.CO;2.

Seiler, A., Evensen, G., Skjervheim, J.-A., Hove, J., & Vabø, J. G. (2009). Advanced reservoir management workflow using an EnKF based assisted history matching method. In *SPE Conference Paper.* https://doi.org/10.2118/118906-MS.

Skauvold, J., Eidsvik, J., Van Leeuwen, P. J., & Amezcua, J. (2019). A revised implicit equal-weights particle filter. *Quarterly Journal of the Royal Meteorological Society.* https://doi.org/10. 1002/qj.3506.

Skjervheim, J.-A., Evensen, G., Aanonsen, S. I., Ruud, B. O., & Johansen, T. A. (2007). Incorporating 4D seismic data in reservoir simulation models using ensemble Kalman filter. *SPE Journal, 12*(3), 282–292. https://doi.org/10.2118/95789-PA.

Skjervheim, J.-A., Evensen, G., Hove, J., & Vabø, J. G. (2011). An ensemble smoother for assisted history matching. In *SPE Conference Paper.* https://doi.org/10.2118/141929-MS.

Snyder, C., Bengtsson, T., Bickel, P., & Anderson, J. L. (2008). Obstacles to high-dimensional particle filtering. *Monthly Weather Review, 136,* 4629–4640. https://doi.org/10.1175/2008MWR2529. 1.

Snyder, C., Bengtsson, T., & Morzfeld, M. (2015). Performance bounds for particle filters using the optimal proposal. *Monthly Weather Review, 143.* https://doi.org/10.1175/MWR-D-15-0144.1.

Soares, R. V., Luo, X., Evensen, G., & Bahkta, T. (2021). Handling big models and big datasets in history matching problems through local analysis. *SPE Journal, 26*(2), 973–992. https://doi.org/ 10.2118/204221-PA.

Souopgui, I., Ngodock, H., Carrier, M., & Smith, S. (2017). A comparison of two preconditioner algorithms within the representer-based four-dimensional variational data assimilation system for the navy coastal ocean model. *Journal of Operational Oceanography, 10*(2), 127–134. https:// doi.org/10.1080/1755876X.2017.1306376.

Spantini, A., Baptista, R., & Marzouk, M. (2019). *Coupling Techniques for Nonlinear Ensemble Filtering.*

Stordal, A. S., Moraes, R. J., Raanes, P. N., & Evensen, G. (2021). p-kernel Stein variational gradient descent for data assimilation and history matching. *Mathematical Geosciences, 53,* 375–393. https://doi.org/10.1007/s11004-021-09937-x.

Stuart, A. M. (2010). Inverse problems: A Bayesian perspective. *Acta Numerica, 19,* 451–559. https://doi.org/10.1017/S0962492910000061.

Talagrand, O., & Courtier, P. (1987). Variational assimilation of meteorological observations with the adjoint vorticity equation. I: Theory. *Quarterly Journal of the Royal Meteorological Society, 113,* 1311–1328. https://doi.org/10.1002/qj.49711347812.

Tamang, S. K., Ebtehaj, A., Van Leeuwen, P. J., Zou, D., & Lerman, G. (2021). Ensemble Riemannian data assimilation over the Wasserstein space. *Nonlinear Processes in Geophysics.* https:// doi.org/10.5194/npg-28-295-2021.

Tamang, S. K., Ebtehaj, A., Van Leeuwen, P. J., Lerman, G., & Foufoula-Georgiou, E. (2022). Ensemble Riemannian data assimilation for high-dimensional nonlinear dynamics. *Nonlinear Processes in Geophysics.*

Tarantola, A. (1987). *Inverse Problem Theory: Methods for Data Fitting and Model Parameter Estimation.* Elsevier. ISBN 0-444-42765-1.

Tarantola, A. (2005). *Inverse Problem Theory and Methods for Model Parameter Estimation.* SIAM. ISBN 10-0-89871-572-5. https://doi.org/10.1137/1.9780898717921.

Thacker, W. C. (1988). Fitting models to inadequate data by enforcing spatial and temporal smoothness. *Journal of Geophysical Research, 93*(C9):10655–10665. https://doi.org/10.1029/ JC093iC09p10655.

Thacker, W. C., & Long, R. B. (1988). Fitting dynamics to data. *Journal of Geophysical Research, 93*(C2), 1227–1240. https://doi.org/10.1029/JC093iC02p01227.

van Hinsbergen, C. P. I. J., Schreiter, T., Zuurbier, F. S., van Lint, J. W. C., & van Zuylen, H. J. (2012). Localized extended Kalman filter for scalable real-time traffic state estimation. *IEEE Transactions on Intelligent Transportation Systems, 13*(1), 385–394. https://doi.org/10.1109/TITS.2011. 2175728.

Van Leeuwen, P. J. (1999). The time-mean circulation in the Agulhas region determined with the ensemble smoother. *Journal of Geophysical Research Oceans, 104.* https://doi.org/10.1029/ 1998JC900012.

Van Leeuwen, P. J. (2001). An ensemble smoother with error estimates. *Monthly Weather Review, 129,* 709–728. https://doi.org/10.1175/1520-0493(2001)129<0709:AESWEE>2.0.CO;2.

Van Leeuwen, P. J. (2003a). A variance-minimizing filter for nonlinear dynamics. *Monthly Weather Review, 131.* https://doi.org/10.1175/1520-0493(2003)131<2071:AVFFLA>2.0.CO;2.

Van Leeuwen, P. J. (2003b). Nonlinear ensemble data assimilation for the ocean. In *Recent Developments in Data Assimilation for Atmosphere and Ocean*, ECMWF Seminar, September 8–12, 2003, Reading, United Kingdom (pp. 265–286). https://www.ecmwf.int/node/13249.

Van Leeuwen, P. J. (2009). Particle filtering in geophysical systems. *Monthly Weather Review, 137,* 4089–4114. https://doi.org/10.1175/2009MWR2835.1.

Van Leeuwen, P. J. (2010). Nonlinear data assimilation in geosciences: An extremely efficient particle filter. *Quarterly Journal of the Royal Meteorological Society, 136,* 1991–1999. https:// doi.org/10.1002/qj.699.

Van Leeuwen, P. J. (2011). Efficient nonlinear data-assimilation in geophysical fluid dynamics. *Computers and Fluids, 46,* 52–58. https://doi.org/10.1016/j.compfluid.2010.11.011.

Van Leeuwen, P. J. (2015). Representation errors and retrievals in linear and nonlinear data assimilation. *Quarterly Journal of the Royal Meteorological Society, 141,* 1612–1623. https://doi.org/ 10.1002/qj.2464.

Van Leeuwen, P. J. (2019). Nonlocal observations and information transfer in data assimilation. *Frontiers in Applied Mathematics and Statistics, 26.* https://doi.org/10.3389/fams.2019.00048.

Van Leeuwen, P. J. (2020). A consistent interpretation of the stochastic version of the Ensemble Kalman Filter. *Quarterly Journal of the Royal Meteorological Society.* https://doi.org/10.1002/ qj.3819.

Van Leeuwen, P. J., & Ades, M. (2013). Efficient fully nonlinear data assimilation for geophysical fluid dynamics. *Computers and Geosciences.* https://doi.org/10.1016/j.cageo.2012.04.015.

Van Leeuwen, P. J., & Evensen, G. (1996). Data assimilation and inverse methods in terms of a probabilistic formulation. *Monthly Weather Review, 124.* https://doi.org/10.1175/1520-0493(1996)124<2898:DAAIMI>2.0.CO;2.

Van Leeuwen, P. J., Cheng, Y., & Reich, S. (2015). Nonlinear Data Assimilation. In *Frontiers in Applied Dynamical Systems: Reviews and Tutorials* (Vol. 2). Springer. ISBN 978-3-319-18347-3. https://doi.org/10.1007/978-3-319-18347-3.

Van Leeuwen, P. J., Künsch, H. R., Nerger, L., Potthast, R., & Reich, S. (2019). Particle filters for high-dimensional geoscience applications: A review. *Quarterly Journal of the Royal Meteorological Society.* https://doi.org/10.1002/qj.3551.

Vetra-Carvalho, S., Van Leeuwen, P. J., Nerger, L., Barth, A., Altaf, U., Brasseur, P., Kirchgessner, P., & Beckers, J.-M. (2018). State-of-the-art stochastic data assimilation methods for high-dimensional non-Gaussian problems. *Tellus Series A, 70.* https://doi.org/10.1080/16000870. 2018.1445364.

Villani, C. (2008). *Optimal Transport: Old and New.* Springer Science & Business Media. ISBN 978-3-540-71050-9. https://doi.org/10.1007/978-3-540-71050-9.

Vossepoel, F. C., & Behringer, D. W. (2000). Impact of sea level assimilation on salinity variability in the western equatorial Pacific. *Journal of Physical Oceanography, 30,* 1706–1721. https://doi. org/10.1175/1520-0485(2000)030<1706:IOSLAO>2.0.CO;2.

Vossepoel, F. C., & Van Leeuwen, P. J. (2007). Parameter estimation using a particle method: Inferring mixing coefficients from sea level observations. *Monthly Weather Review, 135*(3). https://doi.org/10.1175/MWR3328.1.

Vossepoel, F. C., Weaver, A. T., Vialard, J., & Delecluse, P. (2004). Adjustment of near-equatorial wind stress with four-dimensional variational data assimilation in a model of the pacific ocean. *Monthly Weather Review, 132,* 2070–2083. https://doi.org/10.1175/1520-0493(2004)132<2070:AONWSW>2.0.CO;2.

Wang, X., & Bishop, C. H. (2003). A comparison of breeding and ensemble transform Kalman filter ensemble forecast schemes. *Journal of the Atmospheric Sciences, 60,* 1140–1158. https://doi.org/10.1175/1520-0469(2003)060<1140:ACOBAE>2.0.CO;2.

Wang, Y., & Papageorgiou, M. (2005). Real-time freeway traffic state estimation based on extended Kalman filter: A general approach. *Transportation Research Part B: Methodological, 39*(2), 141–167. ISSN 0191-2615. https://doi.org/10.1016/j.trb.2004.03.003; https://www.sciencedirect.com/science/article/pii/S0191261504000438.

Weaver, A. T., Vialard, J., & Anderson, D. L. T. (2003). Three- and four-dimensional variational assimilation in a general circulation model of the tropical Pacific Ocean. Part 1: Formulation, internal diagnostics and consistency checks. *Monthly Weather Review, 131,* 1360–1378. https://doi.org/10.1175/1520-0493(2003)131<1360:TAFVAW>2.0.CO;2.

Weaver, A. T., Deltel, C., Machu, E., Ricci, S., & Daget, N. (2005). A multivariate balance operator for variational ocean data assimilation. *Quarterly Journal of the Royal Meteorological Society, 131,* 3605–3625. https://doi.org/10.1256/qj.05.119.

Whitaker, J. S., & Hamill, T. M. (2002). Ensemble data assimilation without perturbed observations. *Monthly Weather Review, 130,* 1913–1924. https://doi.org/10.1175/1520-0493(2002)130<1913:EDAWPO>2.0.CO;2.

Wikipedia. Heat kernel. https://en.wikipedia.org/wiki/Heat_kernel.

Xie, J., Bertino, L., Counillon, F., Lisæter, K. A., & Sakov, P. (2017). Quality assessment of the TOPAZ4 reanalysis in the Arctic over the period 1991–2013. *Ocean Science, 13,* 123–144. https://doi.org/10.5194/os-13-123-2017.

Xie, X., Van Lint, H., & Verbraeck, A. (2018). A generic data assimilation framework for vehicle trajectory reconstruction on signalized urban arterials using particle filters. *Transportation Research Part C: Emerging Technologies, 92,* 364–391. https://doi.org/10.1016/j.trc.2018.05.009.

Yu, L., & O'Brien, J. J. (1991). Variational estimation of the wind stress drag coefficient and the oceanic eddy viscosity profile. *Journal of Physical Oceanography, 21,* 709–719. https://doi.org/10.1175/1520-0485(1991)021<0709:VEOTWS>2.0.CO;2.

Yu, L., & O'Brien, J. J. (1992). On the initial condition in parameter estimation. *Journal of Physical Oceanography, 22,* 1361–1364. https://doi.org/10.1175/1520-0485(1992)022<1361:OTICIP>2.0.CO;2.

Zhu, M., Van Leeuwen, P. J., & Amezcua, J. (2016). Implicit equal-weights particle filter. *Quarterly Journal of the Royal Meteorological Society.* https://doi.org/10.1002/qj.2784.

Zupanski, M. (1993). Regional four-dimensional variational data assimilation in a quasi-operational forecasting environment. *Monthly Weather Review, 121*(8), 2396–2408. https://doi.org/10.1175/1520-0493(1993)121<2396:RFDVDA>2.0.CO;2.

Author Index

Index

© The Editor(s) (if applicable) and The Author(s) 2022
G. Evensen et al., *Data Assimilation Fundamentals*,
Springer Textbooks in Earth Sciences, Geography and Environment,
https://doi.org/10.1007/978-3-030-96709-3

Printed in the United States
by Baker & Taylor Publisher Services